Pioneers Over Jordan

Pioneers Over Jordan

THE FRONTIER OF SETTLEMENT IN TRANSJORDAN, 1850–1914

Raouf Sa'd Abujaber

I.B.Tauris & Co. Ltd
Publishers
London

Published in 1989 by
I.B.Tauris & Co Ltd
110 Gloucester Avenue
London NW1 8JA

British Library Cataloguing in Publication Data.

Abujaber, Raouf Sa'd
Pioneers over Jordan: the frontier of
settlement in Transjordan, 1850–1914
1. Jordan, Rural regions. Agricultural industries,
history
I. Title
338.1'095695

ISBN 1–85043–116–7

Printed and bound in Great Britain by
Redwood Burn Limited, Trowbridge, Wiltshire

To Mireille with love
— Abu Ziad

Contents

Maps

Foreword

The changing relations between nomadic pastoralists and sedentary cultivators of the land form one of the main themes in the history of the countries lying inland from the eastern and southern coasts of the Mediterranean Sea. There are some regions of steppe and desert where vegetation and water are so limited that nomadic pasturage of livestock is the only possible use of land; there are others so well watered and sheltered that they have always been the home of settled agriculture. Most regions, however, can be used for either purpose, and the balance between those who wish to use it for agriculture and those who need its water and produce for their flocks is a precarious and shifting one.

It is easy to explain changes in the balance in terms of some immemorial hostility between 'the desert and the sown', but in reality the relationship between those who follow the two ways of life is more complex than this would imply. Each needs the other, but each tries to move the balance in his own favour. Where it lies at any time will depend upon many factors: on the strength of governments, which is normally used to support the cultivators, producers of foodstuffs and materials for the city and revenues for rulers and armies; on the demands of markets, both local and distant, for produce of different kinds; and perhaps, in the very long term, on changes of climate.

In our own times, the balance has shifted decisively in favour of the settled life – for a variety of reasons. The power and range of governments have grown enormously; expanding populations demand more food; the coming of the internal combustion engine has destroyed the market for animals of transport, the main product of the fully nomadic tribesmen; the expansion of cities and their industries has provided alternative means of livelihood for villagers and nomads alike. So complete has been the change that it is sometimes difficult to grasp how recent it has been. In the lands lying east of the Jordan, the expansion of agriculture has taken place in little more than a century, that of government in a still shorter space of time.

Dr Raouf Abujaber has the rare advantage of being able to describe and explain this process of change on the basis both of written sources and of memories: his own, and those of older members of his family and other Jordanians whom he has interviewed. He gives us a living picture of the first farmers, caught between the demands of the nomadic tribesmen and those of the government; building their houses and villages amidst the ruins of ancient settlements; ploughing the soil and sowing seeds in land which had not been used in this way for centuries; and producing grain not only for their own needs but to feed the growing population on the other side of the Jordan. His book is not only a work of research, it is also a testimony to the skill, enterprise and courage of the pioneers over the River Jordan.

Albert Hourani
October 1988

Preface

Transjordan was for many centuries the fighting arena between the nomadic tribes who settled in its eastern areas and those who were pushing northwards from the south and the east. It witnessed during the second half of the nineteenth century a great improvement in both social and agricultural standards. Agricultural development and population increase, being so closely connected, came about after 1840 when the Egyptian forces that had occupied Syria for ten years withdrew and Ottoman governmental influence spread gradually southwards from Damascus. The settled population, for centuries mistreated by the nomadic tribes to whom they were forced to pay *khawa* (tribute), was at long last feeling the benefits of stability and security under a Turkish government, which, although corrupt and inefficient, offered a great improvement in many respects over the previous decades.

People in the area of study were also to witness important changes in their way of life between 1860 and 1880. These were brought about by the advance eastwards of villagers and farmers, the interest of beduin shaykhs in the acquisition of agricultural land, and the arrival of relatively large bodies of new settlers. The new settlers were the Circassians who came from the Caucasus by way of the Balkans and the three Christian tribes who immigrated from the unsettled area of al-Karak.

The main aim in writing this book was to put on record the different historical, social and economic factors that played such an important role in the lives of the settled and nomadic populations alike. To achieve this aim, the first section presents an account of the geography of the area and its climate, with a special interst in rainfall, followed by a short history of events during the nineteenth century. The second section is concerned with various aspects of agriculture since, with animal husbandry, this was the main activity in the area. In the last section, six case studies of settlements are treated. Although different in many ways, these settlements had one important common aim, which was to start a new life for the members of their communities; in spite of many difficulties, they were successful.

Sources for the study were found in Istanbul, Jerusalem, Nablus, Damascus, al-Salt, and London. Books, travel reports, archives, consular records, and registers of religious courts were studied and referred to. The documents of al-Yaduda, the Abujaber estate, from the end of the nineteenth and the beginning of the twentieth centuries were also of great help, especially regarding agricultural activity. These documents, and the stories and memories related in the beduin encampments of Transjordan, provided much of the original information.

Acknowledgements

For many years I have been fascinated by the pioneering spirit of the early farmers who spread out in the eastern districts of Transjordan during the second half of the nineteenth century. The hardships they encountered in their new life were so many and great that I felt a compelling desire to put on record the interesting story of their life in the frontier of settlement. I also endeavoured to write about the different aspects of their adventurous life and the great importance they attached to agriculture in their new villages and estates.

The work involved in the research and writing of this book has been for me most rewarding; my efforts, however, would have been far less successful had it not been for the continuous assistance that I have received from so many learned and generous people. It is therefore a great pleasure for me to register my indebtedness and to express my gratitude to them all.

At Oxford I was fortunate to receive the most valuable help of two special friends, Mr Albert Hourani and Mr Norman Lewis. My sincere thanks go to my two supervisors, Dr Derek Hopwood and Dr John Wilkinson; and to Mr C. Gordon Smith, Mr Philip Stewart and Mrs Margaret Owen, as well as to all the other kind friends at the Middle East Centre, St Antony's College, the School of Geography, the Oriental Institute and the Bodleian Library.

I also gratefully acknowledge the assistance given me by my friends Dr ʻAbd al-Karim Gharayba and Dr Abdal-ʻAziz al-Duri of the University of Jordan; Dr ʻAbd al-Karim Rafiq of the University of Damascus; the Revd Dr Carney Gavin of Harvard University; Dr Wolf-Dieter Hütteroth of Erlangen University; and Dr Butrus Abu Manneh of the University of Haifa.

During the years of research I have interviewed over a hundred knowledgeable people; had access to the documents and records of many families and institutions; and made use of the services of many libraries and archives in London, Istanbul, Jerusalem, Damascus, Nablus, Nazareth, Amman, al-Salt, and Ma'an. A list of the ladies and gentlemen who so kindly helped me in all these places would be too long to enumerate here but I wish to extend to each one of them sincere and grateful thanks. I wish also to thank Miss Paula Kaliss, Miss Layla Odeh, Mr Adel Hamdan and Miss Basima Abujaber, who have given so much of their time to the preparation and typing of the manuscript in its different stages, and to Mr Yusuf 'Ubayed for his help with the maps.

A special tribute is extended to my wife Mireille for her understanding and continuous encouragement. Together with my daughter Basima and sons Ziad and Marwan they made life easier for me during the years of work in Oxford and elsewhere.

A Note on Transliteration and Dates

The fact that there were different dialects in the area of study during the nineteenth century has induced me to keep the system of transliterating Arabic words and proper names as simple as possible. Generally these have been rendered according to the system adopted by the *International Journal of Middle East Studies*. In a few instances family names have been written in the form adhered to by the family itself. An example is of course my own family name which technically speaking should have been Abu Jabir; Abujaber was retained since it has been in use for the last hundred years. Words where the spelling has become generally accepted have been kept as such in the text. There is no elision of the definite article and *ta marbuta* has been omitted, even in construct. Turkish words such as *akçe*, *bey*, *effendi*, *rumeli*, and *werko* have been kept in the form that is commonly used and accepted.

Geographical names have been generally applied according to the *Index Gazetteer* and the maps published by the Department of Lands and Surveys of the Hashemite Kingdom of Jordan. In many instances, however, some alterations have been found absolutely necessary.

Dates in this study have always been given as they appear in the documents referred to; practically all of them were dates of the Hijra calendar starting from the Prophet Muhammad's migration from Makka on 1 Muharram I H/16 July/622 AD. The dates for the Gregorian calendar are given after conversion. In this endeavour I have been greatly helped by two reference works: G. S. P. Freeman-Grenville, *The Muslim and Christian Calendars* (London, 1963), and Faik Reşit Unat, *Hicri tarihleri, Miladi Tarihi, Cevirme Kilavuzu* (Ankara, 1974). As to the few instances when the dates were given following the Mali (Rumi) calendar, it was found less cumbersome to convert them in advance and only give these dates in Hijra and Gregorian years.

PART ONE
Environment and History

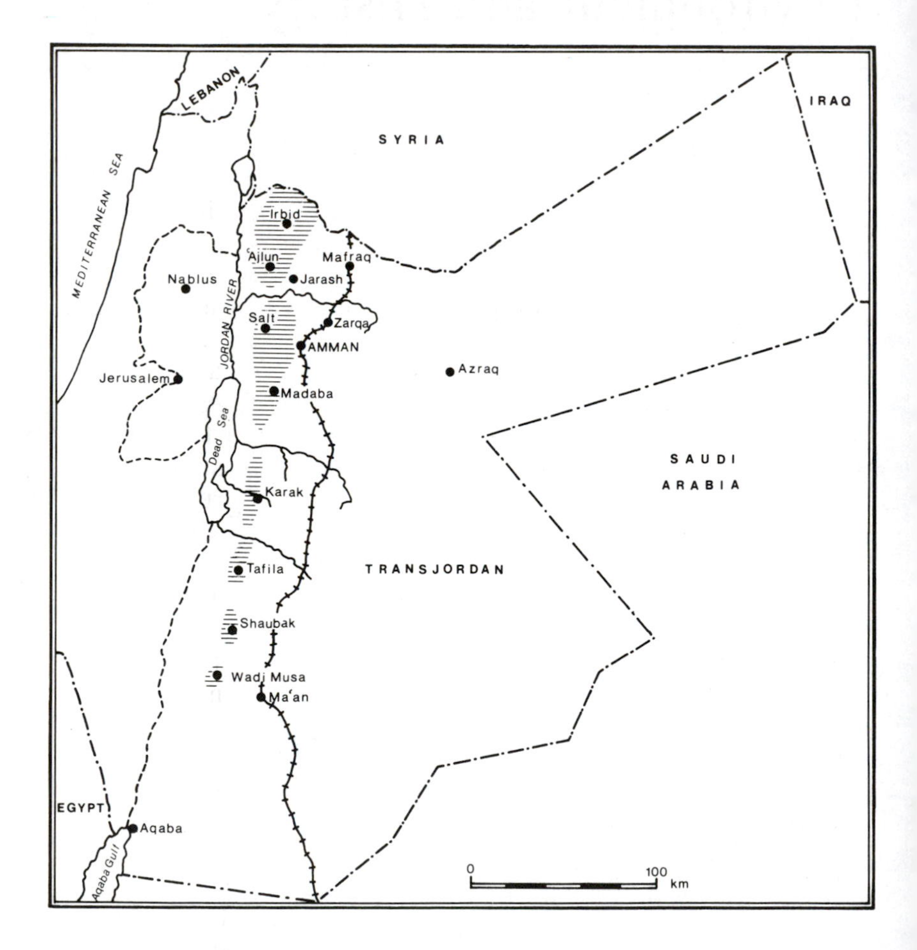

Map 1.1 The situation of Transjordan (shaded sections show cultivated areas during the 1880s)

1 Environmental Conditions of Transjordan

This work is concerned with the study of the nineteenth-century eastern frontiers of settlement in the area that was, prior to the war in Palestine in 1948, known as Transjordan. Its boundaries in the north are the Yarmuk River and a line that runs about halfway between Dar'a in southern Syria and al-Ramtha, the most northern large settlement in Transjordan. It lies to the east of the River Jordan, the Wadi 'Araba and the Gulf of 'Aqaba. In the south is the Hijaz and further to the south-east is Najd. In the south-east and the east the area stretches into the north Arabian and Syrian deserts, to the frontiers of Saudi Arabia and Iraq. Its length from north to south is nearly 400 kilometres and its width varies from 150 kilometres in the narrower parts to over 380 kilometres in the section reaching towards Iraq. The total area is about 96,000 square kilometres of which no less than 80 per cent is steppe and desert. The population at the end of the nineteenth century is not known but a very rough estimate puts the figure at around 150,000, of whom nearly 100,000 were city and village dwellers and the rest nomadic tribesmen of the Transjordanian countryside.

In ancient times Transjordan was divided into the kingdoms of Edom, Moab, and Ammon. During Roman times, it formed the outlying provinces of the Empire. Under the Byzantines, the area flourished since most of the cities of the Decapolis were situated there. The Arab conquest in 636 AD saw the whole area, with Palestine, divided into two adminstrative units: Jund al-Urdunn in the north and Jund

Falastin in the south. Under the Mamluks, al-Karak was the capital of their kingdom of al-Karak, but after the Ottoman conquest in 1516 the region was reduced to a mere outlying section forming the southern part of the *wilaya* of Damascus. This last period of 400 years was indeed one of neglect and hardship; the whole area suffered the consequences of instability and depopulation. At the end of the first world war, Transjordan came temporarily under British administration. Later, both Palestine and Transjordan were entrusted to the British by the League of Nations as mandates. The emirate of Transjordan came into existence during that period; it became fully independent in 1946 as the Hashemite kingdom of Transjordan.

With so many names and changes, it is not easy to decide which name to apply to the study area of this book. At first 'South-eastern Bilad al-Sham' seemed to be the most apt, since Bilad al-Sham, or Greater Syria, has historically been considered to denote the larger entity covering the area between the Taurus mountains in the north and the Red Sea in the south. Presently, the area under study forms the south-eastern part of greater Syria as a whole. After some consideration, however, it was thought more in line with present-day developments to use the name Transjordan, as it will be easier for the reader to associate this term with literature and maps of the first half of the twentieth century (see map 1.1).

The geography of Transjordan has been well documented since 1812 when the Swiss traveller, John Lewis Burckhardt, visited the country and aroused interest in its history and heritage by rediscovering Petra of the Nabateans and writing *Travels in Syria and the Holy Land* (1822). Among other leading works on the subject are the *Historical Geography of the Holy Land* (1894) by George Adam Smith and *The Jordan Valley and Petra* (1905) by William Libbey and Franklin E. Hoskins.

The term Transjordan was used by William, Archbishop of Tyre, in his *History of the Crusades*. The words 'trans Iordanem' in the original Latin version were, however, rendered by his translators as 'region beyond the Jordan'. It may also have been used in earlier times, but the first known reference to it during the nineteenth century was made by Dr H. B. Tristram in 1872 when he wrote 'With the Shaykhs of the Transjordanic Tribes' in his book, *The Land of Moab*.[1] The word Transjordanic was again used a few years later in 1877 in the book of Selah Merrill, *East of Jordan*.[2] Gottlieb Schumacher, the German traveller, used the phrase 'Trans-Jordan countries' during the course of a railway survey of the high plateaux of Hauran and Jawlan in 1886.[3] The name was used in 1905 by the two geographers, Libbey and Hoskins. They used the words, 'Mysterious unbroken skyline', to describe the 'Trans-Jordanic' mountain ridge when seen from the west. Evidently these and others who were interested in Transjordan for various reasons were becoming familiar with the name or were already inclined to use it in their writings about the area. Later, at the end of the first world war, the name Transjordan came into more popular usage and was adopted as the recognized name of the new state which was set up in Amman on 15 May 1923 under the leadership of His Highness Prince 'Abd Allah.[4]

Maps of the region during the period under study generally referred to the different parts of the area by their original and later Turkish administrative names such as 'Ajlun, al-Salt, al-Balqa', al-Karak, al-Shawbak, and Ma'an. However, George Adam Smith found it suitable, when he designed and edited his *Atlas of the Historical Geography of the Holy Land*, which was published in 1915, to apply the word Transjordan to the whole area east of the River Jordan and the Dead Sea.

Two maps from other studies are also felt to be of significance for a study of Transjordan. The first is a map of the irrigation of the Jordanian Valley, prepared in January 1914 by G. Franjia, the Lebanese engineer who also prepared reports on irrigation in the Turkish district of Quniya.[5] The drop between the Hula region at sea-level and the Dead Sea at 432 metres below sea-level, within a distance of only 184 kilometres, has had a great effect on agriculture and living conditions, especially health, in the whole region of al-Ghawr (the Jordan Valley depression). Communication between the two banks of the Jordan was severely impeded by the necessity for any traveller to cross this deep gorge. It also rendered transport and exchange of goods more costly.

The second map (see map 1.2) shows the major catchments in Transjordan and their areas together with isohyets of estimated average annual rainfall (in millimetres) for the period 1901–30. The map's significance lies in the fact that it clearly defines the 'four magnificent rifts' (*wadis*) that traverse the country from east to west and cleave it into four divisions.[6]

The northern and western boundaries of the area included in this research have been naturally defined. In the south, Ma'an, an important station on the pilgrimage road, administered by the Hijaz government since the end of the first world war, was brought under the administration of the government of Transjordan during 1925.[7] In the east, the boundaries are undefined and they expanded and contracted in line with the extension of agriculture in the pioneer fringe area of Badiya al-Sham. This area has been referred to by writers during the last hundred years as the lands of 'the desert and the sown'.[8] A few centuries earlier, Amman was referred to as being on the 'sword of the desert' by the geographer al-Muqaddasi.[9] For him, it represented the boundary between the settled agricultural areas and Badiya al-Sham where nomadic life was dominant. As late as 1926, the report of His Britannic Majesty's Government to the Council of the League of Nations on the administration of Palestine and Transjordan for the year 1925 recognized the situation prevailing in these eastern areas. It confirmed that 'there are large areas of undefined and undelimited wasteland hitherto uncultivated owing to lack of public security. Some are now being revived under the present more settled regime particularly as closer control is exercised over the nomadic beduin.'[10] It is important to state that the process of settlement which was being revived in 1926 was in operation as early as the 1860s. Settlers from the villages around al-Salt and 'Ajlun had by then started their advance to the east, and the area taken over for agricultural purposes was in certain districts over 40 kilometres in width. The frontiers of settlement had begun to expand through a continuous process that is still in operation towards the end of the twentieth century. With the growth of population, whether natural or through an influx of Palestinians, it is not surprising that the demand for more land is continually increasing.

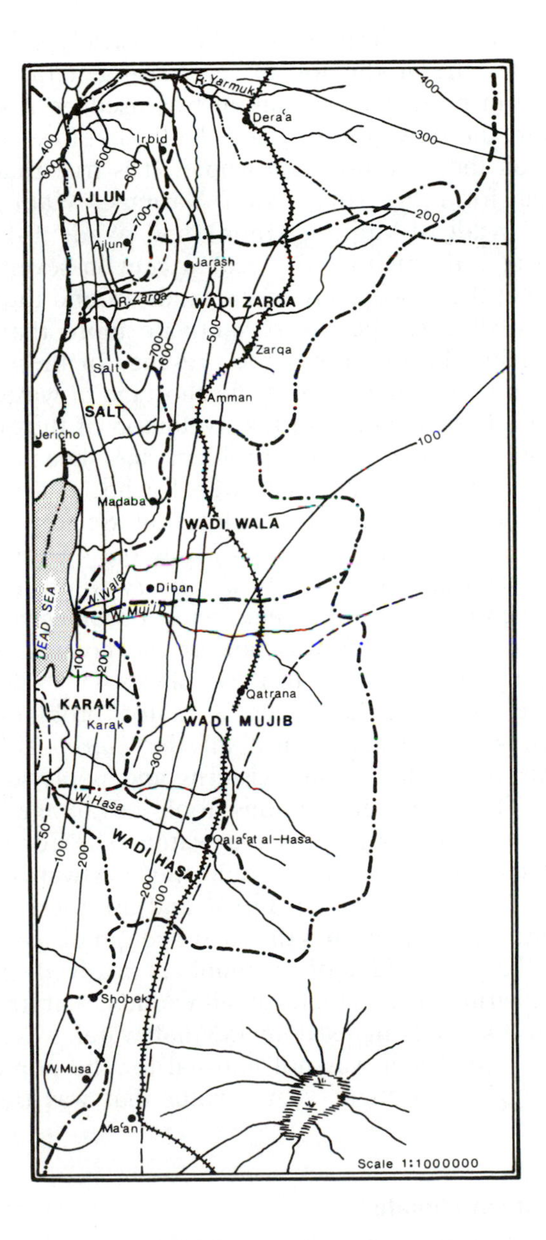

Map 1.2 Transjordan's catchments and average rainfall, 1901–30
Source: Ionides, *Water Resources*, figure 9, p. 26.

Agriculture in the area of study was basically a dry-farming operation for the production of cereals in general and wheat and barley in particular. Irrigation was used on a very small scale near the few springs and had therefore very little influence on the economy as a whole. It is true that the four wadis that form the rifts, Wadi Yarmuk, Wadi al-Zarqa' (Jabbok), Wadi al-Mujib (Arnon) and Wadi al-Hasa, had running water throughout the year except in severe drought seasons. Yet the topography of the lands around them did not make it possible for people to cultivate more than the few irrigable fields. Furthermore, the presence of these wadis or rivers was no guarantee against the shortage of water that has been such a basic element in the livelihood of the Jordanians all through the ages. This problem has been very well described by the traveller Selah Merrill who, as archaeologist of the American Palestine Exploration Society placed in charge of exploration work in 1875, wrote: 'Sometimes we suffered from want of water. In a country where it does not rain for seven months in the year, water becomes very scarce. The want of good water was one of the greatest hardships attending my life and work in these places.'[11]

Shortage of water has been a major problem for the population since ancient times; all the *khirab* (old ruined settlements) have their wells, cisterns and pools to attest to this fact. The older generations took advantage of every possibility to collect and preserve water. The Jordanians of the nineteenth century, when venturing eastwards to settle new lands, had only to clean, repair and use these thousands of wells and water reservoirs that had been in disuse for many centuries. To give an idea of the numbers involved, the author confirms hearing from elders at al-Yaduda that there were more than 300 wells in its domain, that water was stored in these wells and that it was not unusual for the population to drink, during the drought years, water that was three years old.[12]

Transjordanian climate

The climate of Transjordan can be divided into three main types according to the topography of the land, which has distinctly marked longitudinal zones in spite of its relative narrowness in certain parts. The most western zone is the Ghawr or Jordan Valley depression where the weather is very hot in summer and warm in winter, with usually little rainfall. The second zone is the middle hilly region known as al-Shafa (plateau) which enjoys a pleasant climate of the Mediterranean type. This is characterized by a hot, dry summer and a cool, wet winter, with two short transitional periods of spring and autumn in between. Summer usually starts in May and winter around mid-November. The third zone is the steppe or the desert which lies east of the Hijaz railway in the eastern part of the country. This zone forms a part of the Syrian desert and is characterized by its hot summer and cold winter with very little precipitation.

The rainfall over al-Shafa, or the hilly region, is typical of areas that have the Mediterranean type of climate throughout the year. The comparatively high rainfall during an annual wet season that covers about half the year enables dry farming to be carried out in these regions as long as the rains do not fail. This last possibility is a fact of life in Transjordanian existence and the unreliability of rains was, during the nineteenth century, the biggest hazard for dry farming. Failure of rainfall in one year or over a few successive years meant not only bad crops or no crops at all, but also hardship on all levels. Animals sufferd because of lack of pasture and fodder, while people, besides having less water for drinking, were subjected to loss of income from both agriculture and animal husbandry. Unfortunately, there are no records of these bad years of the nineteenth century, but a clear picture of the consequences of rainfall unreliability is provided by an interesting chart compiled by A. A. El-Sherbini (see figure 1.1). This demonstrates how both yield and production have been directly related to rainfall.

To illustrate the regime imposed by this type of climate, it was necessary to gather information about rainfall monthly totals by rainfall years, and not by calendar years, in order to obtain a true picture of rainfall during the rainy season of every year. The eight stations chosen for the exercise and appearing in figure 1.2 are those that had, and continue to have, the largest concentrations of population in Transjordan.

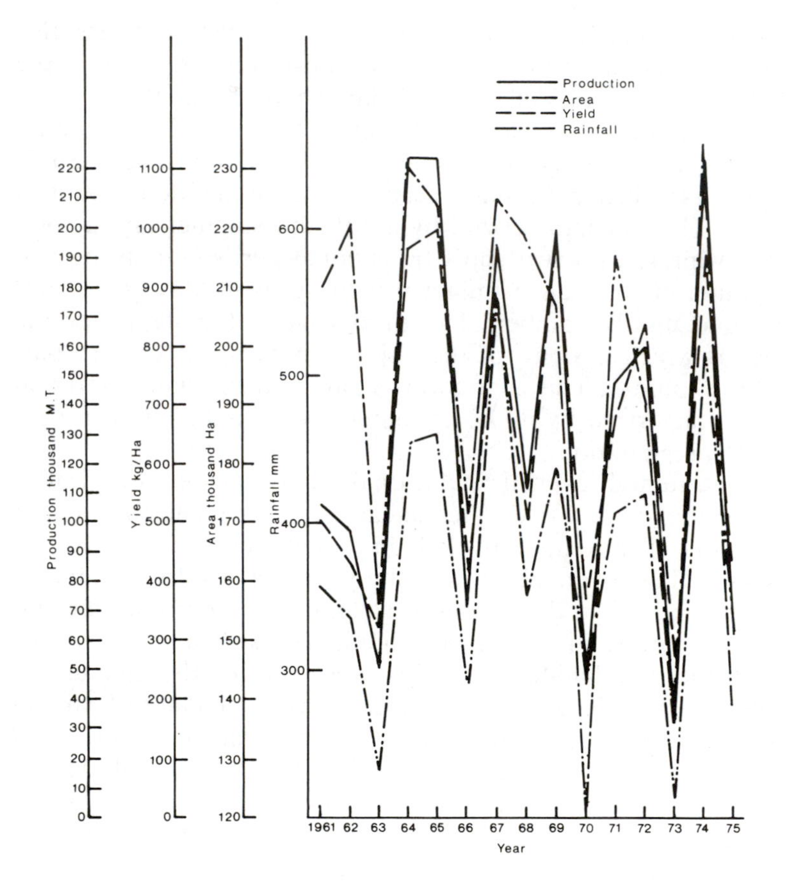

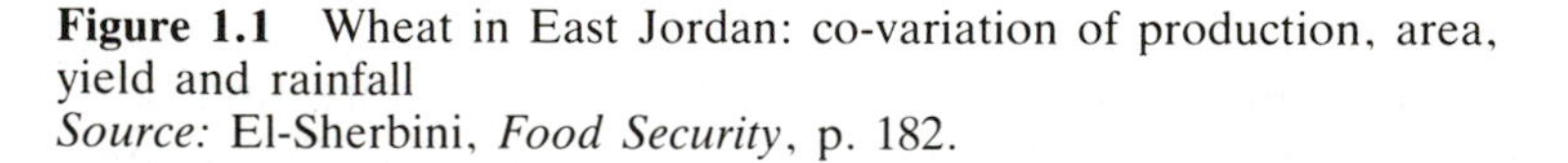

Figure 1.1 Wheat in East Jordan: co-variation of production, area, yield and rainfall
Source: El-Sherbini, *Food Security*, p. 182.

They were also well fitted to the purposes of this study in their geographical locations. They are all situated in al-Shafa and the agricultural areas around all of them have been the most important in the economic development of the area. Furthermore, the stations included are those which had reliable records for more than ten years during the period 1923–65. Amman airport had data for the complete period, due to the establishment in 1921 of a British Royal Air Force base at

Marka, a suburb of Amman. In 1922 their meteorological services were already in operation and thereafter information was available until the British withdrawal in 1958 when the Jordanian Meteorological Department took over the task. The long-term data from Amman airport for 1923–65 were taken by the Meteorological Department as a basis and data from the other stations were correlated with this long-term period. The way this was done is outlined by them as follows:

> The correlation of the mean rainfall amount was made on graphs by the method relationship. The mean rainfall amounts for stations with incomplete records were derived from ratios on the basis of the trend featuring variations of the given precipitation. The diverse distribution of precipitation is attributed to the complications of the relief of Jordan.

This reference to Jordan, in the 1968 report published by the Meteorological Department just after the 1967 war, includes all of Transjordan and also the West Bank, which remains occupied by Israel.

Fig. 1.2 will give an idea of the frequency of rainfall in the different months of the 'rainfall year' which was considered, for the purpose of this study, to start on 1 June and end on 31 May. The pattern of distribution of rainfall through the year is very consistent. As a result of this annual pattern it is understandable why farmers in Transjordan had, and continue to have, an agricultural system that is applied consistently from year to year. The farmers depend heavily on the rains and their scope for adopting new techniques is therefore limited within each season. Variations in the rainfall from month to month are naturally important and make all the difference between a bad year and a less bad or an acceptable year. The total quantity of rain in a rainfall year is, however, by far the most important factor. This rainfall total affects not only cereal and fruit farming activity, but also has a great influence on water resources, whether stored in wells and pools or flowing at a steady rate from fountains and springs. Springs are directly linked to the rising or falling of water tables and shortages of rainfall affect them adversely, although the effect on wells and pools is felt more immediately.

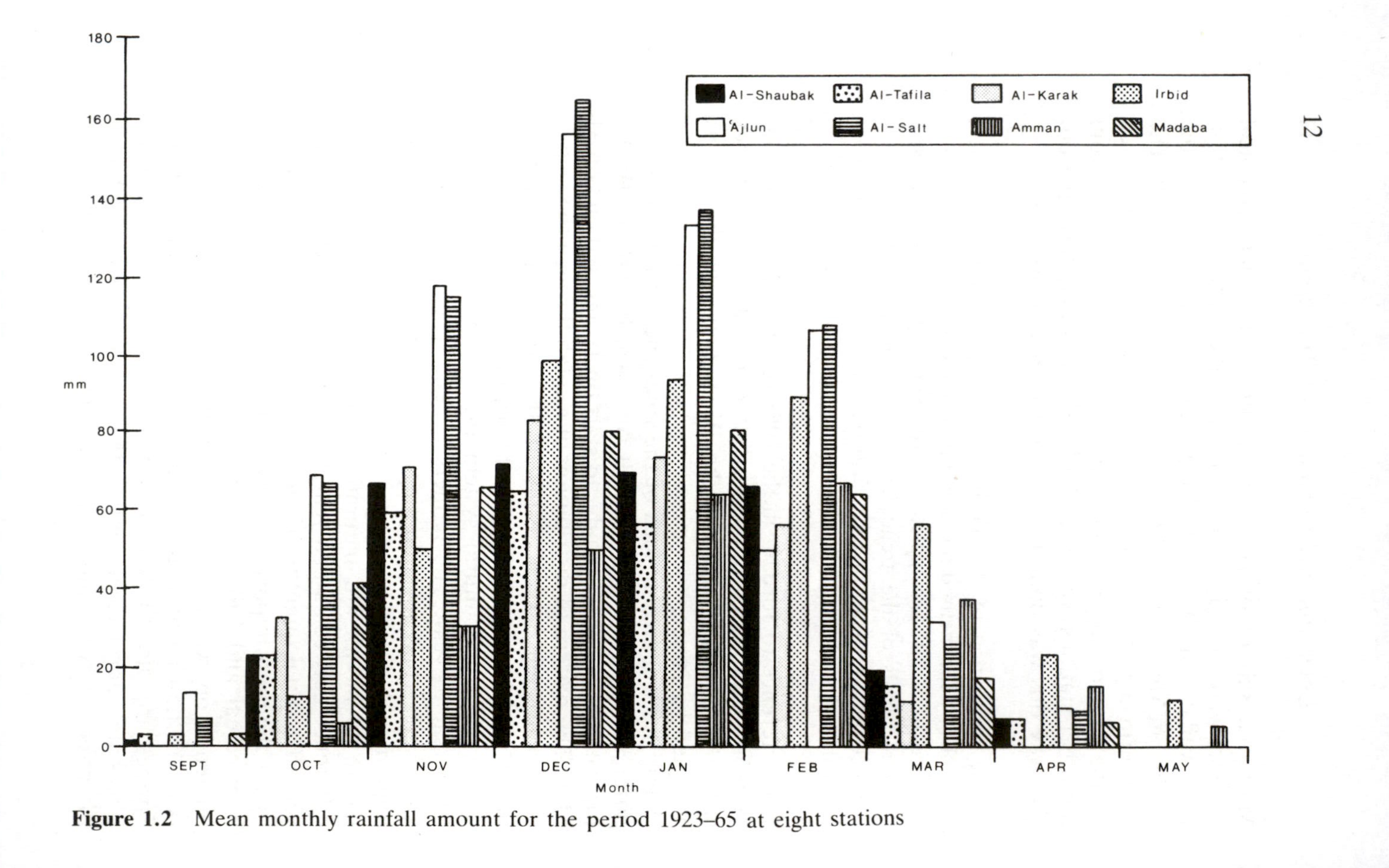

Figure 1.2 Mean monthly rainfall amount for the period 1923–65 at eight stations

The quantity of rainfall that the country receives during a 'rainfall year' is therefore of very great importance. This basic fact made it necessary to find the rainfall figures for years during the period of study. Unfortunately such records are simply non-existent.[13] The Ottoman authorities, as far as it has been possible to ascertain, did not make any meteorological observations at any of their three administrative centres at 'Ajlun, al-Salt, or al-Karak. It was not possible, therefore, to produce actual rainfall figures for any of the years of the nineteenth century relevant here.[14] However, it is on record that the Ottomans established the first rainfall station in Transjordan at Ma'an in 1901. Observations, we are informed, were made during the construction of the Hijaz railway in 1902, but ceased between 1905 and 1934.[15] Sadly, the records of the station at Ma'an, in spite of an extensive search, have not been located. Probably they will never be found, as between 1916 and 1918 Ma'an was in a war zone that witnessed action between the forces of the Arab revolt (supported by the British) and the Ottomans. The Ma'an railway station, where the meteorological station was most probably placed, was attacked twice before it was captured. Thomas E. Lawrence, who participated in the campaign, described what happened to the station after the second attempt succeeded:

> Through the haze of a fierce battle a file of Turks waving white things rose out of their main trenches in a dejected fashion. The force converged wildly on the station and I gained the station hill. A minute later, with a howl, the beduin were upon the maddest looting of their history.[16]

It is unlikely that any records survived this onslaught.

Under the circumstances, a system had to be devised to obtain information about Transjordan's rainfall during the nineteenth century. Fortunately, rainfall records in Palestine were begun in January 1846 by Dr McGowan, the physician to the English Mission Hospital in the Old City of Jerusalem.[17] Another series of daily observations of rainfall was begun in 1861 by Dr Chaplin in the Old City, under the auspices of the Palestine Exploration Fund. The work was later supervised by

Table 1.1A Rainfall in Jerusalem per month 1861–1901

	1861	1862	1863	1864	1865	1866	1867	1868	1869	1870	1871	1872	1873	1874	1875	1876	1877	1878	1879	1880	1881
Rain in inches																					
January	9.66	12.41	9.11	6.89	4.54	5.06	9.23	3.57	7.72	1.21	2.94	3.11	0.13	8.43	6.79	3.42	1.60	14.39	0.98	6.00	1.28
February	6.50	2.27	2.40	1.50	5.08	3.18	6.07	10.93	3.27	0.69	4.42	5.25	6.03	7.22	4.09	4.11	9.75	11.49	2.27	4.01	4.43
March	2.40	0.63	3.70	1.08	0.42	3.46	2.14	3.29	1.95	3.99	6.75	1.43	1.94	10.02	10.52	2.27	0.89	2.35	7.52	5.64	4.36
April	0.32	1.00	2.11	1.65	0.77	0.29	2.01	1.93	2.36	3.72	1.10	0.42	0.89	0.13	1.04	1.97	0.21	0.51	1.52	2.07	2.21
May	0.48	0.00	0.00	0.00	0.37	0.00	0.73	0.14	0.40	0.00	0.19	0.11	0.01	0.00	0.23	0.35	0.00	0.65	0.00	0.10	0.07
June	0.00	0.00	0.00	0.00	0.00	0.00	0.00	0.00	0.00	0.00	0.00	0.00	0.00	0.00	0.00	0.00	0.00	0.00	0.00	0.00	0.00
July	0.00	0.00	0.00	0.00	0.00	0.00	0.00	0.00	0.00	0.00	0.00	0.00	0.00	0.00	0.00	0.00	0.00	0.00	0.00	0.00	0.00
August	0.00	0.00	0.00	0.00	0.00	0.00	0.00	0.00	0.00	0.00	0.00	0.00	0.00	0.00	0.00	0.00	0.00	0.00	0.00	0.00	0.00
September	0.00	0.00	0.00	0.09	0.00	0.00	0.00	0.00	0.27	0.00	0.00	0.00	0.00	0.00	0.03	0.00	0.00	0.79	0.00	0.00	0.00
October	0.00	0.00	1.90	0.00	0.00	1.75	0.00	0.00	0.00	2.29	1.58	0.31	0.01	0.00	0.00	0.03	2.18	0.00	0.82	0.40	0.00
November	0.18	2.96	0.19	2.65	1.56	1.84	2.24	1.19	1.47	0.01	0.10	3.39	4.41	2.51	1.12	1.69	5.02	0.03	0.69	0.86	2.43
December	7.76	2.59	7.13	1.65	5.45	2.97	6.98	8.05	1.17	1.45	6.49	6.24	9.30	1.44	3.19	0.49	7.35	3.00	4.24	13.00	1.72
Means	27.30	21.86	26.54	15.51	18.19	18.55	29.42	29.10	18.61	13.39	23.57	20.26	22.72	29.75	27.01	14.41	26.00	32.21	18.04	32.11	16.50

	1882	1883	1884	1885	1886	1887	1888	1889	1890	1891	1892	1893	1894	1895	1896	1897	1898	1899	1900	1901	Mean of 41 yrs
Rain in inches																					
January	3.08	10.93	6.09	7.79	6.55	12.45	4.63	6.13	11.59	10.23	7.42	7.54	4.80	0.90	9.61	14.46	4.40	6.46	2.50	7.42	6.41
February	12.59	3.79	8.26	2.90	9.51	4.16	1.25	0.83	4.18	6.22	4.09	2.12	6.54	3.07	8.65	7.18	6.19	3.30	10.72	0.15	5.11
March	0.97	5.74	3.75	5.47	5.09	3.78	2.03	3.21	1.87	3.38	1.73	12.35	8.45	5.91	5.50	8.18	7.29	3.21	1.85	1.24	4.10
April	3.65	0.35	2.08	6.52	1.34	0.85	4.74	0.74	4.41	0.25	1.58	0.93	1.94	1.84	2.14	0.00	0.36	1.29	0.14	0.23	1.55
May	0.57	0.00	0.62	0.24	0.43	1.25	0.23	0.00	0.00	0.35	1.04	0.06	0.07	0.12	0.42	0.28	0.00	0.00	0.08	0.99	0.26
June	0.00	0.00	0.00	0.08	0.00	0.00	0.20	0.00	0.00	0.00	0.00	0.00	0.00	0.00	0.00	0.00	0.00	0.00	0.00	0.00	0.00
July	0.00	0.00	0.00	0.00	0.00	0.00	0.00	0.00	0.00	0.00	0.00	0.00	0.00	0.00	0.00	0.00	0.00	0.00	0.00	0.00	0.00
August	0.00	0.00	0.00	0.00	0.00	0.00	0.00	0.00	0.00	0.00	0.00	0.00	0.00	0.00	0.00	0.00	0.00	0.00	0.00	0.00	0.00
September	0.00	0.00	0.00	0.00	0.00	0.00	0.02	0.00	0.00	0.00	0.00	0.00	0.00	0.00	0.00	0.00	0.00	0.00	0.00	0.00	0.00
October	0.07	0.31	0.06	0.07	0.45	0.00	0.32	0.00	0.07	0.40	0.03	0.11	0.00	0.41	0.04	0.78	0.00	0.35	0.10	0.28	0.37
November	0.80	7.59	1.08	0.13	5.03	0.60	7.99	0.57	3.48	2.80	6.64	0.60	6.87	3.73	2.08	3.32	3.83	1.47	0.49	1.69	2.37
December	4.99	3.21	2.02	6.27	3.31	6.72	16.40	2.06	9.83	11.09	8.70	6.83	6.71	7.24	4.46	7.42	6.60	6.35	5.32	5.42	5.67
Means	26.72	31.92	23.96	29.47	31.69	29.81	37.79	13.56	35.51	34.72	31.23	30.54	35.38	23.25	32.90	41.62	28.66	22.43	21.20	17.42	25.87

Table 1.1B Rainfall days in Jerusalem per month 1861–1901

	1861	1862	1863	1864	1865	1866	1867	1868	1869	1870	1871	1872	1873	1874	1875	1876	1877	1878	1879	1880	1881
Number of days of rain																					
January	14	11	10	8	7	14	11	9	15	9	9	6	4	15	14	7	9	14	6	15	3
February	7	7	7	5	8	9	12	18	12	1	11	17	10	12	12	10	13	13	6	12	12
March	5	3	8	4	5	9	8	7	4	9	16	17	11	20	14	7	5	7	17	7	10
April	1	4	7	6	3	4	3	13	8	13	3	4	2	3	4	8	3	2	3	6	8
May	4	0	0	0	3	0	5	1	2	0	1	3	1	0	1	4	0	2	0	1	2
June	0	0	0	0	0	0	0	0	0	0	0	0	0	0	0	0	0	0	0	0	0
July	0	0	0	0	0	0	0	0	0	0	0	0	0	0	0	0	0	0	0	0	0
August	0	0	0	0	0	0	0	0	0	0	0	0	0	0	0	0	0	0	0	0	0
September	0	0	0	2	0	0	0	0	1	0	0	0	0	0	1	0	0	2	0	0	0
October	0	0	7	0	0	5	0	0	0	4	2	4	1	0	0	1	5	0	3	1	0
November	3	9	1	4	11	8	4	7	6	1	1	7	7	6	6	7	11	1	6	5	5
December	13	7	12	7	11	13	10	13	5	4	13	7	13	6	9	3	13	4	6	15	8
Total	47	41	52	36	48	62	53	68	53	41	56	55	49	62	61	47	59	46	46	62	48

	1882	1883	1884	1885	1886	1887	1888	1889	1890	1891	1892	1893	1894	1895	1896	1897	1898	1899	1900	1901	Mean of 41 years
Number of days of rain																					
January	11	19	10	19	15	12	12	13	15	16	17	12	14	6	19	14	12	14	6	11	12
February	16	13	18	9	10	6	7	4	15	11	11	9	14	7	12	15	8	11	17	3	10
March	4	9	10	11	9	8	6	5	9	9	2	15	8	13	14	11	13	8	11	4	9
April	12	3	3	7	5	2	8	3	8	5	6	5	6	7	6	0	2	5	1	4	5
May	4	0	3	1	5	2	2	0	0	3	5	1	1	1	4	2	0	0	1	4	3
June	0	0	0	1	0	0	1	0	0	0	0	0	0	0	0	0	0	0	0	0	0
July	0	0	0	0	0	0	0	0	0	0	0	0	0	0	0	0	0	0	0	0	0
August	0	0	0	0	0	0	0	0	1	0	0	0	0	0	0	0	0	0	0	0	0
September	0	0	0	0	0	0	0	1	0	0	0	0	0	0	0	0	0	0	0	0	0
October	1	3	1	1	2	0	3	0	1	3	1	3	0	2	2	5	0	3	3	2	2
November	4	11	7	1	9	4	13	5	7	6	12	2	12	10	9	8	13	7	3	8	7
December	11	12	2	8	8	12	13	10	17	15	9	15	10	6	6	17	11	11	11	4	10
Total	63	70	54	58	63	46	65	41	73	68	63	62	65	52	72	72	59	59	53	40	56

James Glaisher until 1903. After the departure of Dr Chaplin early in 1883,[18] a table showing monthly rainfall in inches and another showing rainfall days were published (see table 1.1).

Although there are some inconsistencies in these figures and a variety of instruments were used in obtaining the data, an annual series of rainfall equivalents for Jerusalem, called the 'Old City Series', was produced in 1955. Later the series was rechecked and added to by C. G. Smith of Oxford University. As reproduced in table 1.2 below, figures were thus available for all the 128 rainfall years between 1846/7 and 1974/5 and the average annual rainfall for this period was given as 549 millimetres.[19]

Thus, while rainfall records for the Old City section of Jerusalem are available starting from the year 1846 (table 1.2), no data whatsoever exist for rainfall in Transjordan prior to the year 1937, except for Amman airport where data were collected after 1922.

However, the geographical proximity of the various population centres in Transjordan to Jerusalem readily suggests that there might be some relationship between the amount of rainfall for Jerusalem and that for each location in Transjordan. Such a relationship could form a basis for estimating annual rainfall figures in Transjordan during the 90-year period from 1846 to 1936 for which no rainfall records exist, using the annual rainfall data available for Jerusalem during the same period.

To identify and define any such relationship, actual rainfall data, representing recorded measurements of annual rainfall during rainfall years 1937–8 to 1974–5 for Jerusalem and for each of eight main locations in Transjordan, were obtained. For this 38–year period of recorded rainfall data, a computer was used to perform a linear regression and correlation analysis between the annual rainfall values for Jerusalem and those corresponding to each of the eight locations in Transjordan. This enabled the form and the strength of the relationship between the data sets to be assessed.

The results of this analysis appear in table A of appendix 1. Table B presents a summary of the rainfall statistics for Jerusalem and each of the eight Transjordan locations during the period from 1937 until 1975. It shows the mean annual rainfall value and a measure of the variation (standard deviation) in annual rainfall at each station. It also shows the correlation coefficient for the annual rainfall data sets from each of the Transjordan stations and that from Jerusalem.

The results of this correlation analysis indicate that annual rainfall for the selected stations in Transjordan is, in general, highly positively correlated with that for Jerusalem. For some of these stations (e.g. Amman and al-Salt), this correlation is remarkably high, at 0.93 and 0.86 respectively.

For each of the eight stations in Transjordan, a linear regression equation relating the stations' annual rainfall with that of Jerusalem was thus obtained. When an annual rainfall figure for Jerusalem is used in one of these it produces a derived value for the year's rainfall for the corresponding Transjordan station.

Finally, then, the regression equations resulting from the analysis of the actual rainfall data were used to produce estimates of annual rainfall amounts for the selected eight Transjordan stations for the 90-year period between 1846 and 1936, based upon the Jerusalem figures for the same period. Table C in the appendix lists the annual rainfall values, both actual (from 1937) and estimated (from 1846 to 1936), covering the 128-year period from 1846–7 to 1974–5 for all stations concerned.

To assess the levels of confidence and reliability of the estimates produced for each station, the linear regression model was applied to the period 1937 to 1975 for which records of annual rainfall data exist. For each of the 38 rainfall years involved, the estimated value produced by the linear regression equation was computed and compared with the actual measurement for that year. The difference, representing the possible error in estimates, was expressed as a percentage of the actual value.

Table C includes the details of these results, while Table B includes a measure of the expected relative error of the estimates for each of the eight Transjordan locations.

While inspecting the summary of results in table B, it is worthwhile noting that for all of the selected locations in Transjordan the value of the standard deviation is consistently larger than that for the corresponding standard estimate error.

Table 1.2 Annual rainfall Jerusalem (old city)

Rainfall year	Annual rainfall (mm)	9 year moving mean	Accumulated departure
1846/ 7	477		–72
8	412		–209
9	490		–268
1849/50	560		–257
1	687	528	–119
2	526	546	–142
3	356	588	–335
4	700	605	–184
5	542	600	–191
6	640	586	–100
7	794	590	145
8	636	619	232
9	522	607	205
1859/60	560	605	216
1	577	580	224
2	623	546	298
3	588	547	337
4	523	565	311
5	419	573	181
6	486	547	118
7	643	532	212
8	689	518	352
9	633	514	436
1869/70	318	579	205
1	486	600	142
2	469	575	62
3	481	538	–6
4	1004	588	449
5	676	599	576
6	419	611	446
7	353	634	250
8	1090	647	791
9	409	600	651
1879/80	598	605	700
1	675	622	826
2	598	654	875
3	584	608	910
4	718	610	1079
5	575	625	1105
6	638	608	1194
7	674	624	1319
8	433	631	1203
9	731	643	1385
1889/90	524	649	1360

(Table continued)

Rainfall year	Annual rainfall (mm)	9 year moving mean	Accumulate departure
1	742	636	1553
2	646	647	1650
3	825	687	1926
4	630	676	2007
5	518	676	1976
6	770	649	2197
7	796	615	2444
8	632	580	2527
9	525	585	2503
1899/1900	499	577	2453
1	399	580	2243
2	509	584	2203
3	678	562	2332
4	448	574	2231
5	794	587	2476
6	828	610	2755
7	441	640	2647
8	633	629	2731
9	611	640	2793
1909/10	544	600	2788
1	786	558	3025
2	572	567	3048
3	547	550	3046
4	434	537	2931
5	451	538	2833
6	524	538	2808
7	480	537	2739
8	492	540	2682
9	552	535	2685
1919/20	787	540	2923
1	567	513	2941
2	569	512	2961
3	392	513	2804
4	496	493	2751
5	278	468	2480
6	476	454	2407
7	499	438	2357
8	376	430	2184
9	556	403	2191
1929/30	440	414	2082
1	433	414	1966
2	319	398	1736
3	249	427	1436
4	375	433	1262

(Table continued)

Rainfall year	Annual rainfall (mm)	9 year moving mean	Accumulated departure
5	482	454	1195
6	349	463	995
7	639	483	1085
8	612	537	1148
9	629	576	1228
1939/40	509	573	1188
1	500	617	1139
2	742	598	1332
3	718	562	1501
4	457	551	1409
5	748	579	1608
6	470	578	1529
7	286	523	1266
8	529	518	1246
9	757	523	1454
1949/50	494	500	1399
1	250	485	1100
2	673	515	1224
3	496	529	1171
4	544	483	1166
5	333	475	950
6	562	470	963
7	648	448	1062
8	348	439	861
9	420	404	732
1959/60	210	443	393
1	474	450	318
2	414	416	183
3	225	455	–141
4	687	475	–3
5	628	510	+ 76
6	335	509	–138
7	705	522	+ 18
8	596	557	65
9	522	528	38
1969/70	566	549	–45
1	531	562	–63
2	545		–67
3	425		–191
4	818		+ 78
5	453		–18

Mean over 128 years = 549 mm

This implies that estimates based on linear regression using the Jerusalem figures are significantly superior to expectations based solely on the characteristics of the random distribution of annual rainfall amounts for a given location. In other words, the statistical data confirm that the estimates produced in this work, no matter how inaccurate, are likely to be closer to reality than guesses based upon actual knowledge of the nature of rainfall of each location.[20]

When studying the rainfall figures, including the calculated derived figures of the computer, and comparing them with each other and with all available information about each location at the present, some basic climatic trends become evident. The distribution of rainfall over the area of Transjordan conforms in general to the following principles:

1 Rainfall decreases with increasing distance from the sea. Thus al-Salt receives much more rain than Amman.
2 Rainfall increases with increasing altitude. Thus al-Karak receives more rain than Amman.
3 Rainfall decreases from north to south. Thus Irbid receives more rain than Amman.[21]

The use of a computer in this exercise was helpful in that it rendered the estimated rainfall means for eight locations during the relatively long period of 90 years (1846–1936), 53 of which are in the nineteenth century.[22]

2 Transjordan During the Nineteenth Century

The Ottoman conquest of Bilad al-Sham began four hundred years of occupation when Sultan Selim defeated the Mamluk Sultan Qansawh al-Ghawri at the battle of Marj Dabiq near Aleppo on Sunday 25 Rajab 933 H/24 1516 AD. Transjordan, or south-eastern Bilad al-Sham, fell to the Ottomans soon afterwards, without any fighting, and its people could have felt very little of the new upheaval.

As far as we know, the country was not prosperous and was without any cities of importance. It was far from the centres of political and military activity in the Empire. Unlike the areas around Damascus and Aleppo, it had no claims to economic or military importance and was therefore considered by the new Ottoman administration as a poor outlying district that neither required nor deserved any attention. This attitude on the part of the administration changed during the two months of the annual pilgrimage season when the Hajj route and the stations on it from Damascus to Madina bustled with activity. Once the Hajj was over, the area lapsed again into its predominant condition of neglect and inactivity.

Information available to us about the course of events and the way of life during the first three centuries of Ottoman rule in Transjordan is rather scarce. It is certain, however, that the area bordering on Badiya al-Sham (the Syrian desert) endured a continued state of instability due to a conflict of interests between the nomadic tribes and the Ottoman government, particularly with regard to the Ottoman administration of the Hajj caravan to the holy places. The Ottomans strongly believed that it was their solemn duty to safeguard the pilgrimage. The beduin tribes had equally strong convictions: living on both sides of the route, they maintained that this was their *dira* (domain) and therefore they had a right to the benefits of this yearly venture into their tribal territories. These divergent views were the cause of much friction, especially since the Ottomans did not try to arrange the permanent presence of a real force in the area.

Exhaustive research has not revealed much written history about the first 300 years. Studies such as this, therefore, have to depend on records and documents that are kept in the archives at Instanbul, Damascus and Nablus, as well as on reports of travellers and consuls, whenever available. Ottoman care for records and documents, together with their zeal for tax collection, has fortunately preserved for us an excellent example of fiscal activity as it was at the end of the sixteenth century. This is the *Daftari Mufassal Jadid* (New detailed register) which was compiled in 1005 H/1596–7 AD and thoroughly studied by the two geographers, Wolf-Dieter Hütteroth and Kamal Abdulfattah. The data used in preparing table 2.1 are derived from their detailed published work. The table gives the number of villages, the population estimates and the amount of taxes due from every province in *akçes*, an Ottoman silver coin (called asper by Europeans). Forty akces were equivalent to one Ottoman gold piece during the sixteenth century but by 1730 the value of the coin had fallen, and one gold piece was worth 300 akces; it was finally abandoned at the beginning of the nineteenth century. The final population figure for Transjordan of just under 52,000 people is in line with around 50,000 for Hawran, 40,000 for Jabal Nablus and 42,000 for Jabal al-Quds (Jerusalem). An important aspect of the population distribution in Transjordan during this period is the non-existence of any cities. People lived in small villages or encampments; the largest settlement was ‘Ajlun which had a total of 364 households, or around 1,800 people.

During the nineteenth century, the situation for the population improved. The government in Istanbul, beginning with the reign of Sultan Selim III (1789–1807) and the institution of the *Nizam-i cedid* (new troops), was attempting

Table 2.1 Population and taxes in Transjordan in 1596/7 AD

Province	Fiscal units		Population	Taxes	Main town
	inhabited	empty	estimate	Akçe	or village
Northern part: Qada' Hawran					
Nahiya Kfarat	10	—	905	53,818	Samar
Nahiya Bani Kinana	52	—	6,270	583,483	Hibras
Nahiya Bani Juhma	20	—	4,035	214,304	Irbid
Nahiya Bani 'Atiyya	8	—	1,150	56,028	al-Husn
Nahiya Bani al-A'sar	29	—	5,405	206,351	Kitim
Nahiya al-Batayna	4	—	860	39,900	Al-Ramtha
Total	123	—	18,625	853,884	—
Northern part: Liwa' 'Ajlun					
Nahiya 'Ajlun	18	5	3,420	97,658	'Ajlun
Nahiya Bani 'Ilwan	29	2	4,175	138,776	Jarash
Nahiya al-Kurah	23	—	2,825	174,524	Tubnih
Nahiya al-Ghawr	13	3	1,950	84,453	Dayr 'Ala
Total	83	10	12,370	495,411	—
Middle part: Liwa' al-Balqa'					
Nahiya al-Salt	5	4	4,050	122,000	al-Salt
Beduin Encampments of Bani Sakhr, Smaydat Na'im and Bani Mahdi	—	—	7,120	173,040	—
Total	5	4	11,170	295,040	—
Southern part: Qada' al-Karak wa al-Shawbak					
Nahiya al-Karak	14	—	3,120	134,400	al-Karak
Nahiya Jibal	7	—	1,425	60,000	al-Tafila
Nahiya al-Shawbak	8	—	1,750	51,000	al-Shawbak
Beduins in Nahiya al-Shawbak	—	—	3,425	60,250	—
Total	29	—	9,720	305,650	—
Grand totals					
Northern Part					
Qada' Hawran	123	—	18,625	853,884	
Liwa' 'Ajlun	83	10	12,370	495,411	
Middle Part					
Nahiya al-Salt	5	4	11,170	295,040	
Southern part					
Nahiya al-Karak etc.	29	—	9,720	305,650	
Total	240	14	51,885	1,949,985	

Source: see text.
Note: Total taxes equivalent to 48,750 Ottoman gold pieces.

to introduce a new programme in the system of government, administration, education, and taxation.[1] This trend continued during the reign of Sultan Mahmud II (1808–39), a period described as 'the Turning Point'.[2] However, until the latter part of the nineteenth century, the reforms were barely felt in much of Syria, where the Ottoman authority at the time was weak.

The population had been mistreated for centuries, either by the governors, such as Ahmad Pasha al-Jazzar, wali of Acre, or beduin tribes in the different districts. People therefore migrated from one place to another in an attempt to find a better and more peaceful life. Al-Salt, whose people in March 1806 'are free from every kind of taxation and acknowledge no master',[3] had its share of this population movement in the form of a wave of immigrants, who not only added to the population's numbers but also to their crude skills. About a hundred of them came from Acre and Nazareth.[4] People in al-Salt[5] relate that a certain mason of al-Far family, who migrated from Nazareth around 1798, was commissioned by the local dignitaries to build a water fountain in the midst of the town for a fee to be paid equally by every household. When he completed the work, many of the families did not fulfil their promise to pay. As there was no governmental authority to which al-Far could complain, he found it necessary to take matters into his own hands. He chose a dark night, slipped into the fountain area and with a stone that he had previously carved, he closed the fountain completely. At dawn, the people discovered that they had no water supply any more, and it did not take them long to find out why. All the outstanding debts were settled instantly, and al-Far applied his skill again to bring back running fountain water to the thirsty people of al-Salt.[6]

Instability during the first years of the nineteenth century was not only due to local factors. In 1224 H/1809 AD the Wahhabi thrust northwards was acquiring greater dimensions. Sa'ud Ibn 'Abdul Aziz, the Saudi prince who had already occupied Hijaz and its holy places, led his tribesmen with the declared intention of conquering first the wilaya of Damascus and later the whole of Bilad al-Sham. When the news reached Damascus that Ibn Sa'ud was already near al-Muzayrib in the

Hawran, the wali, Yusuf Kanj Pasha, gathered whatever forces he could and went out to meet the invaders. At the same time he called for assistance from Sulayman Pasha al-'Adil (the Just), successor as wali of Acre to al-Jazzar who had died in 1319 H/1805 AD. Sulayman responded favourably, moved his forces and issued instructions to all the notables in his domain to meet him with their forces in the environs of Tabariyya, the city on the lake of the same name (Tiberias). Among those who answered the call were the tribesmen of Transjordan led by Sa'd al-Qi'dan al-Fayiz,[7] shaykh of one of the groups of the Bani Sakhr, and Ifhayd, shaykh of the second faction. They both gathered their horsemen and joined the campaign at Tabariyya. In the meantime, Sultan Mahmud II (1808–39) had sent a firman to Sulayman Pasha, ordering him to relieve Yusuf Pasha of his post, confiscate his wealth, and put him to death. Somehow, a certain Shaykh Nimr, of the Bani Sakhr, acquired knowledge of this secret. He galloped at night to al-Muzayrib, informed Yusuf Pasha of the affair and rode back without anybody knowing about his mission. Ibn Sa'ud found it more expedient to withdraw since the element of surprise was no longer on his side and the combined forces he had now to fight were much stronger than he had anticipated. Yusuf Pasha, in turn, withdrew to Damascus, where he was besieged by Sulayman Pasha who was intent on executing the orders of the sultan. A battle between the forces of the wilaya of Damascus and the wilaya of Acre took place just outside Damascus, at the village of Darayya. The wilaya of Acre was victorious, and Yusuf Pasha immediately fled to Egypt where he was very well received in Cairo by yet a third govenor of the troubled Ottoman empire, Muhammad 'Ali Pasha.[8]

The importance of the role played by the Transjordanian tribes in the political and military events was once more strongly manifested. Their function, as people of the dira between Hawran and northern Hijaz, was to fight back any beduin incursions into their territory from the south or the east, and at the same time to maintain control of the annual pilgrimage operation and its economic benefits. They would have welcomed more stability in their own domain but the weakness and inconsistency of the central government made this impossible.

The administration in Istanbul arranged matters so that governors did not stay long enough in their posts. They were fleeced before they got their appointments and whatever expenses they incurred were mercilessly collected many times over from the already impoverished people in the *wilayat* (provinces). They were made to fight amongst themselves, which, besides draining the resources of the wilayat, brought havoc and destruction to the countryside. The governors played the same game but on the level of the administration in their own governorships. They encouraged faction against faction and in the Transjordanian area they appeased the big tribes, such as the different branches of the ʿAnaza, and tightened their hold over the smaller ones.[9] It was the people at large who suffered the consequences of all this. The well-known historian Satiʿ al-Husari, in the early twentieth century, gave this description of the times: 'These events were not out of the ordinary in the history of the Ottoman Empire. They were the symptoms and natural outcome of illnesses and diseases that had penetrated the body of the state, causing disintegration in its different structures without sparing a single one.'[10]

The Wahhabi threat, which was averted by the unusually fast reaction of the two governors in Damascus and Acre, however, did not die completely. A year later, Burckhardt confirms a new, but more peaceful, attempt at expansion. Ibn Saʿud is reported as:

> Sending two Wahhabi Tax-gatherers from Madina to al-Karak but they departed without obtaining a single piaster. During their stay, however, tobacco was banished from the guest's room at the Shaykh's house, in conformity with the religious practices of the Wahhabis and the Muslims of Karak showed their adherence to the faith, by going regularly to prayers which few of them were in the habit of doing, the Shaykh excepted.[11]

The shaykh was Yusuf al-Majali who, together with his descendants and clansmen, played an important role in the political life of the area all through the nineteenth and twentieth centuries.

The interested reader can obtain a good picture of general

conditions in the early nineteenth century as they were found by certain famous travellers: the Austrian, M. Seetzen, known as Hakim Musa, in the autumn of 1806; the Swiss, John Lewis Burckhardt, known as Shaykh Ibrahim, in the summer of 1812; and the Englishman, James Silk Buckingham, known as Hadjee 'Abballah in the winter of 1816. Practically contemporaneous with Buckingham, two commanders in the British Royal Navy, the Hon. Charles Leonard Irby and James Mangles, visited Transjordan and recorded their notes and impressions.[12] The notes of the Finnish traveller George August Wallin, known as Shaykh 'Abd al-Wali,[13] are also interesting although his visit took place some 30 years later in the spring of 1845. The population numbers given every now and then were similar to those at the end of the sixteenth century.

All the travellers realized the difficulties of making such trips in countries that did not have any central government. They were all at the mercy of, and subject to the exactions of, the beduins. This often explains the bitterness in some of their remarks and the warmth with which they mention hospitality for its own sake. They were all impressed by the possibilities of a land that had a very small population and prayed for the peace that would allow it to prosper. Although some of their remarks are personal and prejudiced, the present generation is nevertheless grateful that these records have been written.

Generally, conditions were most unfavourable for the development of agriculture. Almost the entire population, whose livelihood depended on different farming activities, was caught between two evils. On the one hand, they had to live under a weak and corrupt administration, while on the other they had to coexist with a harsh and exploitative beduin presence. Although both the administration and the beduins despised farming, they were ready to take advantage of the impoverished people engaged in it. They not only mistreated farmers but were also intent on exploiting them to the last degree. No wonder therefore that villages were very small and only a few families lived in each of them. On the whole, villages were larger and more abundant in the mountainous areas than in the open country, as the mountains afforded better protection against frequent raids of beduin horsemen.

The Egyptian occupation of Bilad al-Sham between 1831 and 1841, in itself, was another sign of the general internal weakness in the Ottoman Empire. Muhammad 'Ali Pasha, the wali of Egypt, and his son Ibrahim conquered all of Syria and for ten years consolidated four predominantly Arab provinces, Egypt, Damascus, Aleppo and Sayda, into a single political state. Egyptian rule for Syria in general meant a better and more progressive administration,[14] but for Transjordan it did not really bring much change. The Egyptian presence, at most times, was nominal, probably because the countryside did not have the potential normally required by conquerors. It did not have the population density that would provide a good number of conscripts for the army; not did it offer any opportunity for collection of taxes on a worthwhile scale. Furthermore, the lines of strategic and commercial communications and supply that were so important under the Mamluks during the fourteenth and fifteen centuries[15] were now replaced by the sea routes and the coastal roads. The improved performance of the Egyptian fleet, which was reorganized in 1829 on the basis of European standards,[16] meant faster and less costly transportation of army reinforcements and supplies after the original major expedition had made its thrust into Syria.[17] This situation greatly reduced the importance of south-eastern Bilad al-Sham and led to its neglect by the new administration.

The records of the Egyptian campaign that are available to us do not mention the prevailing conditions in Transjordan, nor do they report in any detail the events or the military activity during the first stages of the Egyptian presence. It is probably safe to assume that the state of anarchy, tribal domination and limited interest in agriculture continued to prevail. Some useful information is derived from the minutes of the meeting held by the Majlis al-Shura (consultative council) in Damascus, on 24 Safar 1248 H/23 July 1832 AD. These state clearly that the beduins were, as before, uncontrollable, and confirmed that:

> the beduins were committing very unjust acts and mistreating people in Hawran, Irbid, and 'Ajlun, through plunder, taking by force and highway robbery. What lies behind this arrogance

> and greed is the absence of any cavalry forces in Bilad al-Sham. The *mutasal'im* (governor) of Hawran is good for nothing and is more greedy than the beduins.[18]

The lists for the distribution of Muhammad 'Ali's forces in Syria during that period clearly reveal that there were no Egyptian forces anywhere south of Damascus or east of the Jordan River.[19] Transjordan was, for another period, governing itself; there was little change from the conditions of the eighteenth century. The tribes that were committing these outrages against the settled population were mainly Transjordanian tribes, such as the 'Adwan, Bani Sakhr, Sardiya and Sirhan. The Wuld 'Ali[20] and Ruwala[21] tribes of 'Anaza, who also undoubtedly undertook similar exploits, used to frequent the Transjordanian districts in spring and summer, at their will.

In the middle and southern parts of the country, a similar state of affairs prevailed within even the settled community. Egyptian dispatches state that the shaykhs of Nablus and Jinin visited Ibrahim Pasha in Haifa on 4 Jumada al-Awal 1247 H/20 November 1831 AD to present to him their submission. During the visit they petitioned the Pasha, informing him that Ahmad Bay Tuqan, who was hated by the population and had fled from Nablus, was residing at the castle of al-Salt.[22] The Egyptian *siraskir* (commander) evidently did not know that al-Salt, governed by its local shaykhs, was willing to have as guests dignitaries who were out of favour with the Egyptian rulers. The shaykhs of Nablus and Jinin, under the leadership of the 'Abd al-Hadi family, therefore found it necessary to mention the matter in the hope that the Pasha would remove their competitor from a stronghold that was only ten hours' ride from their city.

Ibrahim Pasha was kept so busy by uprisings and revolts in Palestine and other parts of Syria that he did not find the time or any urgent need to subdue Transjordan. However, his commanders found it necessary every now and then to send punitive raids, like the one during Shawal 1248 H/February 1833, when Ma'ajun Agha and a detachment of cavalry 'were prevented by snow from attacking the Bani Sakhr whereupon they plundered some of their animals'.[23]

Later in that year, however, Ibrahim was confronted with the Jabal Nablus uprising under the leadership of Qasim al-Ahmad who, after fleeing to Hebron, took refuge in al-Karak. The Pasha followed him and laid siege to the town which resisted for 17 days.[24] On entering al-Karak after a bloody battle he announced it free for all, and for five days his soldiers plundered everything in it. He dispatched his forces behind the withdrawing people of al-Karak as well as the tent-dwellers in the area around it and plundered from them thousands of sheep and hundreds of camels and animals of burden.[25] The shaykh, Isma'il al-Majali, who fled to the south was taken prisoner and executed in Jerusalem.[26] The remainder of the Christian community, 740 people in all, who had previously conveyed to the Pasha their intention not to participate in the fighting, were allowed to leave the town. Seeing how poor and practically naked they were, the Pasha ordered that twenty camel-loads of wheat, which had been looted from the town, be given to them.[27] They migrated to the environs of Jerusalem where many of them died of hunger and cold. Those who remained alive went back to al-Karak in 1841, after the withdrawal of the Egyptian forces to Egypt.[28]

At the orders of Ibrahim Pasha, al-Karak was razed to the ground.[29] When the mighty fortification and the city wall resisted their picks and crowbars, the Egyptian army used barrels of gunpowder to blow them up. The noise of the explosion was so great that it was heard in Jerusalem, which is about 50 miles from al-Karak.[30] Trees were uprooted and burnt and cisterns and wells were damaged or rendered useless.[31]

Marching northwards, the Pasha attacked the Bani Sakhr in Zizia and, after subduing them, demolished the castle. He continued his advance against al-Salt, destroying its castle and damaging parts of the town. Dhiyab, shaykh of al-'Adwan, was exiled to Homs where he stayed until 1841.[32] The expedition, which had started as a pursuit of a Palestinian chief, became a nightmare of destruction for the population of the middle and southern parts of Transjordan. Many of the people deserted their villages and migrated to other areas and the drop in agricultural activity all over the country must have been enormous.

The Egyptian period in Transjordan was therefore one of suffering, poverty and instability, even though Ibrahim Pasha himself was a lover of agriculture and gave it his personal attention.[33] He succeeded elsewhere in Syria in settling people in rebuilt villages and farms, stopped the *khawa* (tribute money) and threatened to punish severely all those who acted against the public order.[34] In Transjordan, however, people had to wait many years, especially in the al-Karak area, for an acceptable standard of security and stability which would allow agriculture to become again the major occupation.

After ten years, the people of Transjordan were not unhappy to see the end of Egyptian rule. They were even encouraged to participate in armed struggle to expedite the withdrawal and a record of these days of strife can be found in three books of the day. The first, *Narrative of the Late Expedition under the Command of Admiral the Hon. Sir Robert Stopford*, by W. P. Hunder, mentions the officers who were sent into the Syrian countryside, 'to urge every Syrian to rise and harass to the utmost of his power the retreating forces of Ibrahim Pasha who had shown symptoms of evacuating Damascus'. The second, *The War in Syria*, by Commodore Sir Charles Napier, reports how Ibrahim Pasha succeeded by quick movements between al-Muzayrib, Zizia, al-Salt, Al-Karak, and Jericho in holding off the Ottomans and their allies and withdrawing from Transjordan with the bulk of his forces. The third, *Reminiscences of Syria*, by Lieutenant Colonel E. Napier, gives an account of Napier's activity with his 100 irregular horsemen at Umm Qays and also of Count Szecheni, a captain of the Austrian Dragoons and a relation of Count Metternich. Assisted by Captain Lamé of the French Army, the Austrian officer moved his 600 irregulars, who were raised at Tubna in Nahiya al-Kura a few days earlier, and engaged the troops of Ibrahim Pasha at Jarash during the first few days of January 1841. At the same time, similar actions were also taking place in Lebanon, Nablus, Jabal al-Khalil, and Khan Yunis, and officers from the British, French and Austrian armies were helping the Ottomans and enlisting the help of the local populations to bring about the change.

South-eastern Bilad al-Sham did not immediately feel the winds of change that blew over the other provinces in the

Ottoman Empire during the Tanzimat era, which brought reforms that were started by Sultan ʻAbd al-Majid I (1839–61). The countryside was poor, with a small population that paid meagre taxes whenever the governmental administration had the military means to stop beduin oppression of the settled people. The northern part continued to live within the framework of the local *mashyakhat* (chieftainships), basically under beduin domination whether as allies or foes.[35] This situation continued until the late 1840s when the Ottomans tried to extend their control southwards and announced the establishment of the Liwa' of ʻAjlun, to which was attached Irbid and al-Balqa'. Although this administrative addition appeared in the *Salname-yi devlet-i aliyyeh-yi osmaniye* of 1266 H/1849 AD, it does not seem to have been put into effect until 1851, when a governor was appointed and the district came under governmental control. Unfortunately the salnames of 1850 to 1853 are missing from the collection at the Basbakanlik in Istanbul but that of 1854 mentions Liwa' ʻAjlun with only Nahiya al-Kura attached to it. The Salname of 1855 mentions four districts attached to the Liwa', ʻAjlun, Irbid, al-Balqa', and al-Karak, and this, in the absence of any actual Ottoman physical control, must be explained as an indication of the intentions of the wali of Damascus to try to control the countryside. To achieve their aim, the Ottomans resorted to a new drive to assist farmers and at the same time co-operated with the beduin elements that seemed more ready and willing.

Among these beduin factions, there was a group of Maghariba and Hawwara (North African and Egyptian tribesmen) who played an important role in the life of northern Palestine and Transjordan between 1840 and 1870, under the leadership of an outstanding man by the name of ʻAqila Agha. His father, Musa Agha al-Hasi, a member of al-Bara'asa tribe in Burqa, had migrated from Egypt to Gaza and later joined al-Hawwara in Damascus as an *agha* (commander of irregular forces) with 50 horsemen. Towards 1820 he came to Galilee where he married a Turkmani girl; their son ʻAqila was born near Nazareth. In 1847 the Ottoman government appointed him as the commander of 75 irregular horsemen and delegated him to protect the American mission under Lynch. Until his death in 1870 he served the government in many posts, especially as collector of taxes in Galilee and Transjordan. His forces, which reached 500 horsemen on certain occasions, were mainly Egyptian Hawwara but he frequently allied himself to the beduins of the land. On many occasions he revolted against the Ottomans, and in one instance, around 1863, could find no safe refuge except al-Salt, whose people extended to him and his men hospitality of the first degree. For around 50 years, ‘Aqila, his son Quwaytin and grandson Rida, all of whom adopted the title agha, managed to remain as officials of the Ottoman administration. Due to the weakness of that same administration, and by allying themselves with different tribes at different periods, they succeeded in maintaining themselves as a tribal-administrative element, acting in their own interest. However, because they had no blood relationships with any clan, as opposed to the ‘Adwan and al-Fayiz, they were unable to preserve their position of strength when the government took over the direct administration of Transjordan in 1920.[36]

‘Aqila's period is important because it coincided with two very important developments in Bilad al-Sham as a whole. The first was the sad events of the communal strife in 1860 in Lebanon and Damascus, which could have spread to Nazareth and elsewhere had it not been for the wisdom and courage of this outstanding leader. Reports of his stand in these dark moments against fanaticism and vandalism are full of praise.[37] The second was that he witnessed the governmental drive to control the countryside, especially after 1867 when a *qa'im-maqam* (administrative officer of a qada‘) was established in al-Salt. As he felt the change occurring towards the end of his life he must have advised his sons to adapt accordingly; subsequently we hear of them as tax collectors in the service of the local governors, completely lacking the influence that once was that of ‘Aqila Agha.[38]

In May 1869 the British Consul-General in Damascus, Mr W. Wood,[39] and his French colleague, Mr Stein, proceeded to al-Muzayrib where Rashid Pasha,[40] the Governor-General of Syria, had arrived a week previously to meet the caravan from Makka and to settle affairs connected with the Hawran. On their arrival, the Pasha informed them that Fandi al-Fayiz, at the head of 1,500 of the Bani Sakhr, and ‘Ali al-Dhiyab, with about 500 ‘Adwan, had the evening before pillaged al-Ramtha on account of the refusal of its villagers to pay the khawa. This had continued to be exacted in virtue of the ancient custom, although Rashid Pasha had abolished it in his expedition against the beduins in 1867. He complained that the Bani Sakhr, who had the privilege of escorting the Hajj, never ceased giving trouble to the authorities. They invariably used the occasion of the passage of the caravan through al-Balqa’ to make demands to which the authorities were compelled to accede in order to avoid a confrontation which might compromise the safety of the pilgrims. The security of the caravan was rightly regarded as an object of primary importance by the government. The proximity of the Governor-General gave a more serious character to the audacious act of Fandi al-Fayiz; it could not be passed over without considerably weakening the prestige of the government and thereby dispelling the sense of security that the Pasha was endeavouring to establish.

An advance force – consisting of a battalion of infantry on dromedaries carrying Snider rifles;[41] 600 *zaptiehs* (irregular troops); and Muhammad al-Dukhi, shaykh of Wuld ‘Ali, with 800 horsemen – moved southwards the next day. The Pasha, who persuaded Isma‘il al-Atrash to join them with 160 Druzes, invited the two Consuls-General to accompany the expedition, which offered great interest, not so much on account of the military operations, rendered easy by the superiority of the troops sent in advance, as for the opportunity afforded to observe the resources of the country so little known by reason of the difficulty of its access. The hope of making a journey which might be of interest to their respective governments, coupled with the desire to witness the manner in which the various elements rode together for the first time, persuaded the two Consuls-General to accept the invitation on the spot. They started with the reserve force, consisting of 1 battalion of infantry, 1 of cavalry, 60 Druzes, 300 beduins of various tribes of al-Balqa’, as well as two pieces of rifled cannon. By inducing the tribes of al-Balqa’ to espouse the cause of the government, the Pasha aimed not only to obtain over the enemy a moral advantage rendered still greater by the presence of the far-famed Druzes, but also to demonstrate to his superiors in Constantinople a clear proof of his success.

Once the expedition was about to attck the ‘Adwan, their chief, ‘Ali al-Dhiyab, requested *aman* (safe conduct) which was granted on condition of the payment of 25,000 piastres, an amount representing both the expenses of the expedition and the value of the property taken at al-Ramtha. On submitting to the conditions, ‘Ali was allowed to enter the camp. He gave as his reasons for the attack on al-Ramtha that the news of the war between Turkey and Greece, of the continued insurrection of Crete necessitating the withdrawal of troops from Syria and of the wali's departure for Constantinople, had led both him and his ally, Fandi, to the belief that it was the most favourable time for obtaining their respective ends. His object was to rescue his father who was a prisoner at Nablus[42] and Fandi's object was to re-establish his authority by levying the khawa. Though the ‘Adwan surrendered, the Bani Sakhr still held out and sought the alliance of the Hamayda who inhabited the difficult passes between Hisban and al-Karak. On arriving at the valley of al-Wala, the Governor-General established sole control over all supplies of water, thereby forcing the Bani Sakhr either to retreat into the desert where starvation was awaiting them or to fall back on al-Karak where a force was in readiness to give them battle. Soon after, therefore, Fandi sued for the aman which was granted on the condition that he pay 200,000 piastres and give his son as a hostage.[43] The privilege of escorting the caravan was taken away from the Bani Sakhr and given to the Wuld ‘Ali,[44] who only a few days earlier had defeated the Bani Sakhr and had taken from them 2,000 sheep, 300 camels, and 140 oxen.[45]

The Pasha then summoned al-Hamayda, who had given their protection and assistance to the Bani Sakhr to make their submission. On finding themselves for the first time in the presence of a governor, the shaykhs of this tribe openly declared that they were not aware they were under the authority of the sultan, nor that they were inhabiting his territory, as from time immemorial their domain had been in their possession. Nevertheless, they promised henceforth to recognize the government of the sultan and to pay taxes. The Pasha assembled the shaykhs of all the tribes of al-Balqa' on 2 June 1869 and declared to them that if any disturbances occurred no consideration whatsoever would deter him from returning to punish the offenders, who would be compelled to pay the costs of the expedition as well as increased taxes. Before leaving the next day for al-Salt, he entrusted the collection of taxes and fines to 'Aqila Agha who had lately joined the camp with 500 horsemen, and conferred upon the shaykhs of his beduin allies robes of honour.[46] Naturally, he was greatly satisfied with the results of his expedition which, without his knowing, marked the end of the independence of the two tribal confederations in the district of al-Balqa'. Thereafer the shaykhs of Bani Sakhr and al-'Adwan alike found it more expedient to co-operate with the government, even becoming officials in its immediate service.[47]

This heavy blow to the influence of beduin tribes in al-Balqa' was actually preceded by another campaign against al-'Adwan that marked the establishment of governmental authority in al-Salt after an absence that perhaps stretched for more than 150 years.[48] Information about that campaign is contained in a telegram dispatched by the Governor-General of Syria, Rashid Pasha, to the *mutasarrif* (governor of a sanjaq) of Damascus, dated 10 September 1867. It mentions a battle between government troops and the rebellious al-'Adwan at Hajar al-Mansub near Zarqa Ma'in where fifty of the 'Adwan were killed.[49] Rashid Pasha was determined to bring order to south-eastern Bilad al-Sham and he and his successors seem to have succeeded rather well in the two districts of 'Ajlun and al-Balqa'. Security improved a great deal in both areas after 1870 and although fighting and raiding continued among the local population and between them and the beduin tribes, there is no incident on record when government authority was contested.

The arrival of the Circassians in Amman in 1878 and the three Christian tribes in Madaba in 1880 must have also encouraged the trend, already established by the Turks, towards stability and relative security. The newcomers had, as farmers, a basic interest in the quiet that can only be provided by a strong government, which they supported wholeheartedly. The facts about the settlement of these two communities on the frontiers of al-Badiya are most interesting and, together with four other study cases, are discussed at length in their respective chapters.

Virtually unopposed, the entry of Turkish troops into al-Karak city was completed on 23 December 1893. It involved more than one thousand officers and soldiers under the direct command of Husayn Hilmi Pasha. This force was reinforced later and it is possible that Turkish troops in the southern parts of Transjordan numbered four thousand towards the end of 1894 as detachments accompanied officials to other towns in the south. A regular administration was, at long last, established.

A major Turkish achievement in Transjordan was the establishment of the Hijaz railway. The line from Damascus reached Amman at the end of 1902, Ma'an in 1904 and Madina in 1908. But this meant for the Transjordanians and especially for the beduins among them loss of income as the pilgrims caravan no longer existed.[50] Although farming extended to the area adjoining the railway line, people started to realize that they were being called upon to pay higher taxes not only because of government rules but also because of excessive greed among tax collectors as well as corruption among government officials. Tribesmen resented the brutal and unfair treatment they received at the hand of government forces and their shaykhs were unhappy with the confiscation and expropriation of land for different reasons. The general situation was one of dissatisfaction, and in spite of warnings from shaykhs in the different areas the wali of Damascus and his staff do not seem to have taken notice of the impending trouble.

The outbreak came in 1910 when it appeared that preparations were being made to increase taxes, disarm the tribes and conscript men for army service. These rumours, together with the general state of dissatisfaction, led to a series of uprisings along the edge of the desert from Jabal al-Duruz in the north to al-Karak and al-Tafila in the south.[51] Tribesmen tore up railway lines, cut telegraph wires, held up trains and stopped their passengers. They also attacked half a dozen railway stations, and killed some of the employees and army men who were on garrison duty. A number of officials and soldiers were also killed in al-Karak and al-Tafila. However, these actions were neither co-ordinated nor widely supported, and the army had little difficulty in bringing the countryside again under control. Affairs then continued much as before until the outbreak of the first world war in 1914.

PART TWO

Agricultural Activities in Transjordan

3 The Agricultural Cycle

Farming pioneers who spread out into the eastern parts of Transjordan increased in numbers over the years and were already an active community towards the end of the nineteenth century. The boundary of the cultivated areas was being pushed eastwards all the time and *fallahin* (peasants) from the villages of the northern districts, al-Balqa' and al-Karak areas, as well as beduins who turned into farmers,[1] were engaged in agricultural activity.

They all applied a uniform pattern that had been used by their grandfathers in the fields to the west for many centuries. The newly cultivated and sometimes settled areas were only used for dry farming. Olive and fruit trees were rarely planted[2] and therefore the real effort was directed towards the production of cereals in general and wheat in particular, as it fetched a higher price on the market.[3] There were differences between the smaller holdings and the newly acquired large estates, as far as their organizational set-ups were concerned. The fallahin in the already settled villages resorted to the ancient system for cultivation of their *musha'* (common land-ownership system) fields. Generally, every family was limited to a *faddan* or two and the men in the family were the actual *harathin* (sing. *harath*) (ploughmen or farm hands).[4] They co-operated with their cousins and neighbours in the same village and ran their farming at a more or less leisurely pace. This, however, could not be the system in larger estates such as al-Yaduda or its sister farms, each of which had tens of *murabi'iya* (farm hands on yearly contract) and hundreds of animals to care for. Although both were similarly governed by climatic conditions, the larger units had to be better organized and more efficiently run.

The agricultural cycle was the same for both types in its basic elements. It has not been possible to find more than hearsay reports concerning the activities involved in the small farming operations, and it is therefore difficult to give authentic and documented facts about smallholding agriculture during the last decade of the nineteenth century and the early decades of the twentieth. Fortunately this is not the case for al-Yaduda, since some records for the first and second decades of the twentieth century have been preserved. This study has therefore been based on the contents of these records, and although they relate to large-scale farming in principle, it is safe to assume that they applied also to small-scale farming, especially with regard to the basic activities that depended on the climatic conditions such as broadcasting and harvesting.

Broadcasting: the days of hope

Ploughing in these new farms was generally begun towards the end of October when the soil was already ploughed in what became known as *kirab*. If the famer had the resources and the animals he would plough the fields a second time (*thnaya*) or a third time (*thilath*). Fields that were thus serviced and left to rest as fallow for one year were called kirab and were considered fit for the planting of the best crops. They were therefore normally reserved for wheat, the king of all cereals.[5] If they were planted without rest, the ploughed fields were called *malliq*.

Ramy al-bidhar, or the broadcasting of seeds, was normally started at the beginning of November before the rains. All areas sown with seeds prior to the rains were called *'afir*, while those that were sown after the rains were called *riy*. Table 3.1, which gives a record for the season of 1908/9,[6] shows the daily activity of a *rabta* (pl. *rabtat*. work gang) of 18 *faddans* (see appendix 2). Generally, all the work-force followed a general plan regarding attendance at work but, in some instances, climatic conditions imposed certain changes. A rabta sowing

Table 3.1 Daily broadcasting of seeds for 1908/9. One rabta of 18 faddans (14 pairs oxen and four mules) at al-Yaduda, in kayls of 12 sa'a (72 kgs)

Days	Date	Wheat	Barley	Beans	Ni'manih	Lentils	Kirsanih
15	Sat. 15 Nov.						
to	Sat. 29 Nov.	31					
4	Sun. 30 Nov.						
to	Wed. 3 Dec.		28				
3	Thur. 4 Dec.						
to	Sat. 6 Dec.			6	8		
1	Sun. 7 Dec.			14			
1	Mon. 8 Dec.		3				
1	Tues. 9 Dec.		6				
1	Wed. 10 Dec.		No work				
1	Thur. 11 Dec.		No work				
1	Fri. 12 Dec.		5½				
1	Sat. 13 Dec.				10		
1	Sun. 14 Dec.				2	4	
1	Mon. 15 Dec.					5	
1	Tue. 16 Dec.	5					
1	Wed. 17 Dec.		No work				
1	Thur. 18 Dec.	4					
15	Fri. 19 Dec.						
to	Fri 2 Jan.	67½					
1	Sat. 3 Jan.	5					
1	Sun. 4 Jan.		No work				
5	Mon. 5 Jan.						
to	Fri. 9 Jan.	25					
4	Sat. 10 Jan.						
to	Tue. 13 Jan.		24				
1	Wed. 14 Jan.		No work				
1	Thur. 15 Jan.		No work				
1	Fri. 16 Jan.		7				
1	Sat. 17 Jan.		7				
1	Sun. 18 Jan.	5¼					
1	Mon. 19 Jan.	5¼					
1	Tue. 20 Jan.		No work				
1	Wed. 21 Jan.	3½					
1	Thur. 22 Jan.	6					
1	Fri. 23 Jan.	5½					
1	Sat. 24 Jan.	12	(Assistance given by another rabta)				
1	Sun. 25 Jan.	3					
5	Mon. 26 Jan.						
to	Fri. 30 Jan.	25					
1	Sat. 31 Jan.		No work				
1	Sun. 1 Feb.	5					
4	Mon. 2 Feb.						
to	Thur. 5 Feb.		No work				
1	Fri. 6 Feb.	3½					

(Table continued)

Days	Date	Wheat	Barley	Beans	Ni'manih	Lentils	Kirsanih
3	Sat. 7 Feb.						
to	Mon. 9 Feb.		No work				
1	Tue. 10 Feb.						5¼
1	Wed. 11 Feb.						9
1	Thur. 12 Feb.						10¼
1	Fri. 13 Feb.						10
91 days							
Total in Kayls		211½	80½	20	20	9	34½
Total in kg		15,228	5,313	1,440	1,440	648	2,484
Average per Faddan in kg.		846	297	80	80	36	138

Note Ni'manih and kirsanih are two varieties of vetches used for animal feed.

wheat in the western fields of al-Yaduda may have had to stop work because the land was too muddy, while another rabta sowing barley in the eastern fields, 5 kilometres away, may have been able to complete a full day's work. This was the effect of the very local nature of the rainfall.

The quantities sown every day varied in relation to the seed being sown and the time of the season during which the work was being performed. It is apparent that the tempo was slow at the start of the season when sowing was 'afir (average daily 2–3 *kayls* (see appendix 2) per rabta). The pace was faster as the season developed or the rains came (average daily 5 kayls). There were days when assistance came from another rabta, which had probably had more success in meeting its programme and therefore could afford a day of work in fields other than those allotted to it. Records report this happened on 24 January 1909, when 12 kayls were sowed instead of the usual 5. Most probably this was through help from the largest rabta of 24 faddans who joined them in a hard day's work.

Rains were the decisive issue during the sowing season and work was only stopped when they were heavy or the fields were too muddy to be tilled. During the season 1908/9, which started on 15 November 1908 and ended on 13 February 1909, there were 15 days when work could not be performed out of a total of 91 days.

The same trend seems to apply to another rabta of 24 faddans during a different year. The records for the season 1912/13 reveal that the programme which started on 22 November 1912 and ended on 13 February 1913 had 37 days of no work.

As the days of no work were those when it was either raining hard or the fields were too muddy, one can deduce that 1912/13 was a year of heavier rain than 1908/9. The records of 1912/13 stated that on 3 January 1913 the day was spent in a *shiqaq* operation (opening new land to cultivation) and this must have been carried out in the hilly country in the western fields where small plots of land among rocky areas were sowed with barley after the removal by picks of the wild strong weeds and bushes. It is evident that the effort to reclaim arable land did not cease; viewing the smaller fields now, one cannot but appreciate the really hard work undertaken to clear these patches and prepare them for cultivation.

The quantities of seed varied from rabta to rabta and from field to field. The broadcasting by the shaykh and his judgement as to whether to give a certain field *bidhar i'ba* (intensive seeding) or *bidhar dallil* (sparse seeding) were dominant factors. When comparing the seed quantities for the five years under consideration, one discovers that the averages vary. Seeds per faddan in different years varied from each other, the quantities of rainfall and the quality of the fields under cultivation that year playing a most important role.

Table 3.2 illustrates the extent to which quantities, expressed in kilograms per faddan, varied from year to year and rabta to rabta.

The comparison of seed quantities in different years per faddan shows that variations happened even within one locality: a faddan did not have a fixed area of cultivable land. Thus it is safe to assume that the word faddan in the countryside as a whole is used loosely and, although it provides good estimates, cannot be taken as a basis for exact economic research. A faddan had, out of necessity, to have a harath or murabi' for the whole season. The use of the term is therefore helpful in determining roughly the number of the workers in farming communities and thus the numbers of

Table 3.2 Seeds sown by various rabtas in the years 1908–9 to 1912–13 in kilograms per faddan

		Wheat	Barley	Beans	Ni'manih	Lentils	Kirsanih	Sorghum or chickpeas	Sesame
1908–9	Rabta of 18 faddans	846	297	80	80	36	138		
1908–9	Rabta of 18 faddans	1148	290	80	80	36	136		
1909–10	Rabta of 19 faddans	936	379	98.5	106	56.8	155	30	45
1910–11	Rabta of 25 faddans	1172	296	1.3	1.5	1.2	1.8		
1911–12	Rabta of 29 faddans	767	245	60	70	119	189		
1912–13	Two rabtas each of 15 faddans	828	277	79	113	127	163		14
1912–13	Rabta of 24 faddans	881	410	96	143	178	141		

families residing in villages and farms. Wadi al-Sir, a predominantly Circassian village, at the turn of the century ran between 250 and 260 faddans as a whole. The population of that village in that period can therefore be assessed at about 300 families or a total of nearly 1,500 people.[7]

Another interesting point to note is the timing of seed broadcasting of the different products. Barley was sown at the start of the season during November, while vetches, with the exception of *kirsanih*[8], were commonly broadcast during December. In 1908/9 the last four days of the season, 10–13 February, were used for the sowing of kirsanih alone (34.5 kayls) and the last three days of the 1912/13 season were also used for the sowing of kirsanih alone (47 kayls). This is explained by the need to harvest this product before the advent of summer; therefore the crop was allowed the shortest possible time for growth.[9]

Weeding: a pleasant task in spring

It is to be remembered that, when farmers moved eastwards and cultivated land on the fringe of al-Badiya, they had literally to break the soil (*kasr al-ard*). It was very hard work and only strong animals could help the farmers in their activity. In addition to the actual opening of furrows in the solid mass of soil, they also had to remove the plants that had, over hundreds of years, established themselves in the old pastures. Some were so hard to remove that parts of bushes remained in the fields for many years in spite of continuous efforts to remove them. Most known among them were: *Quram* (cat's tail hair grass), *Koleria phleoides* of the gramineae family; *Bilan* (prickly scrubby burnet), *Petorium spinosum* of the gramineae family; *Injil* (Bermuda grass), *Cynodon dactylon* of the gramineae family.[10]

The performance of these tasks, even for the large-scale farmers, seems to have been a continuously difficult operation. Further, these unwanted plant squatters were not alone, as farmers had to contend every year with other types of foreign plants that came with every season. They were called *al-ghariba* (the strangers) and arrived in different ways. Chief among these were *al-ghalath*, or contaminated seeds, such as having barley seeds in a quantity of wheat that was used for broadcasting. Winds brought small seeds of a large variety of wild plants. The droppings of the many animals and birds that frequented the fields after harvest and during the ploughing season contained undigested seeds, and these contributed to the growth of different plants.

Among the wild plants that were more troublesome for established farming units were four that deserve a special mention. They were so rampant that even today one may come across a field of wheat in mid-April with a green base (the colour of wheat stalks and tops) beautifully mixed with the yellow, flowering wild plants that will be then in full bloom. These four are: *Murrar* (three star thistle), *Centurea calcitrapa* of the compositae family; *Khubayza* (cheese weed), *Malva parviflora* of the malvaceae family; *Khurfaysh* (silvery plumless), *Cordus argentatus* of the compositeae family; *Umm udhayna* (bucks horn plantain), *Plantago coronopus* of the plantaginaceae family.[11]

Farmers realized the importance of this additional task and those among them who could afford to remove the foreign plants when they were still green did so without hesitation.[12] The advantages of such extra effort and expenditure were many and obvious. First and foremost the fertility of the soil was preserved for use by the major crop planted in the fields. Furthermore the harvest, which was a completely manual operation, was then an easier task; once the thorny plants dried they became a real nuisance to the bare hands, feet and legs of the workers. There would also be less work during the transport of the harvested stocks, as only those of the original crop would have to be carried to the *baydar* (threshing ground). This is indeed no mean saving as wild plants, if not weeded in time when green, could form up to one third of the harvest's hay-volume[13] in certain unclean fields. In addition, there was also a direct economic advantage accruing from the better price that 'clean' (i.e. uncontaminated with foreign seeds) crops commanded on the market. People who were buying cereals for their own use as well as farmers who were buying for seeding purposes were particular about cleanliness and were willing to pay a higher price for the better grade. Farmer-producers knew very well that it was much easier to prevent contamination of their crops with foreign seeds through using better grade seeds and through weeding than through *gharbala* (sifting) during *baydar* (threshing). When better seeds were used it was much easier and less costly to remove stones and mud pieces from a uniform crop.

Records for this phase of agricultural activity have not been kept and specific information concerning the exact time of its implementation therefore cannot be provided in this study. However, the information available from accounts given by the late Sa'd Abujaber (a large-scale farmer himself until 1964) and from personal experience in the early 1940s furnishes a general outline. Weeding started during the last week of February and ended by mid-April of every year. The tempo of the operation was dictated by the rains and whether the fields were muddy or not. *'Ashabin* (weeders) was the common reference to the group who performed the work; the term in the singular was rarely used. *'Ishaba* was commonly used for the operation as a whole. Numbers of workers in each rabta varied in different fields and for different crops but it is safe to assume that a worker and a half per faddan were required if a good result was to be attained during the six weeks open for the activity. It was therefore not uncommon to find between 30 and 35 workers attached to a rabta of 24 faddans. Among these only one or two would be from the original murabi'iya and they would, in most cases, be performing duty as supervisors; the rest of the murabi'iya would be busy with ploughing the fallow fields and preparing them for the next season.

Among the weeders, women greatly outnumbered men. Generally they were the daughters and wives of the workers on the estate. They found the work agreeable because of its cleaner and lighter nature, especially on sunny days. They had no obligation to report to work every day like the murabi'iya, and if some of them had other things to do they just stayed at home. Boys who were not old enough to become stable boys or shepherds also made some extra money by working as 'ashabin; the only men who joined were those who were not fit for normal work.

Wages in this branch of activity were generally lower than for other daily jobs and varied at the turn of the century between one-half and 3 piastres per worker per day. As such, weeding was not an expensive item on the list of farm expenditure. A liberal estimate would give us the following computation for a whole year's operation at al-Yaduda in these years: an average of 120 'ashabin daily between 25 February and 15 April of every year. Considering that there were days during that period of rain or when fields were muddy, it is reasonable to assume a working season of 42 days in all. The average wage was around 2.5 piastres per worker per day. Thus: 120 'ashabin at 2.5 piastres per day = 300 piastres; and 42 days at 300 piastres = 12,600 piastres. Since a *sa'* (6 kilograms) of wheat cost 3 piastres, the whole operation cost 4,200 sa's or 25,200 kilograms. At the equivalent of 114 piastres to the French gold pound, that meant a total cost per year of 110 French gold pounds, or 126 Turkish gold pounds.

The social aspect of this operation was much appreciated by all those living on the estate. After the long and dreary months of winter, it was good to be out again in the fresh air of the sunny fields. It was a good opportunity for women, girls and boys to make some extra money as well as to gather a good supply of palatable wild herbs that would add to the limited variety of food at home. Knowledgeable old hands at the farm were also certain that these 'ishaba days were popular among the women because they provided them with the opportunity to gather freely and exchange news – a privilege that was not often available in the traditional atmosphere of the village compound.

Harvest: long days of hard work

Work in the farm was continuous, since few jobs had to be performed at the same time. Therefore labour was, during seven or eight months of the year, allotted to the different tasks in a relaxed atmosphere. But this was not the case when the frenzy of harvest started. Then literally every person was mobilized from mid-April to ensure that the crop was removed from the fields to the *bayadir* (sing. *baydar*, threshing ground) in the shortest possible time. This attitude persisted year after year until the 1940s when modern farming methods were introduced. Farmers had always a restless feeling of insecurity as long as their crops were standing in the fields. There was a remote possibility of fire and a more frequent danger of locusts, neither of which they could control. They also worried about thefts at night by marauding parties who not only carried away crops but damaged ten times as many as they took. The most serious worry was reserved, however, for those relatively big waves of slow-moving herds of camels and flocks of sheep and goats that needed pasture and water. Beduin tribes such as the Shararat of Wadi al-Sirhan, different sections of the ‘Anaza, especially the Ruwala[14] and the Wuld ‘Ali, and the different clans of Ahl al-Dira (tribesmen of the domain)[15] converged on the cultivated areas by the start of every May. The pasturage in the Badiya had by then been depleted and its water resources had become dry, while the Jordan Valley, with its bountiful supply of water, was too hot for the animals and unhealthy for fallahin and nomads alike.[16]

The harvest months were those of intensive activity. It was carried out in two forms: *qila‘a* (pulling the plants from the roots) and *hasida* (using the sickle). The first, also called *zahfa* (crawling), was applied always to vetches, lentils, chick-peas and sesame and meant the pulling out of the plants completely from the ground. It was also applied to wheat and barley when the bad years of minimum rainfall did not give the plants the opportunity to grow higher than 30 centimetres. Usually the qila‘a *al-qatani* (harvest of vetches, etc.) started in the first half of May and lasted for about 15 days. Extra hands were hired if necessary, as the barley, the stems of which tended to break easily, had to be harvested right on time to avoid losses. This was in most cases completed by 15 June when the wheat was ripe and ready for the sickle.

On the day set for the start of the wheat harvest, or what is called *fitaha al-manjal* (the opening by the sickle), the whole population of al-Yaduda was already in high spirits. It was a feast day and usually a she-camel or an ox was killed for the party. Harvesters worked at leisure with much showmanship and the rabtat arranged it that all were back in the households by 5.00 p.m. so as to be ready for the dinner at around 6.30. Women who worked in the fields joined in such parties and many of the children flocked to have a meal and attend the *samir* (party) that was held afterwards. The dancing was generally either a *sahja*, which is the beduin semi-circular dance of the men with sometimes one or two girls playing the sword in the middle, or a *dabka*, which was more the dance of the fallahin to the tunes of a *shibaba* (flute). Girls very rarely participated in this dance, which was symbolic of the basic social differences between the beduin and the fallahin. Beduin women had more freedom and clever sword-play by a young beduin woman was considered a sign of accomplishment.

Arrangements in the field were extensive and followed set rules. The shaykh al-rabta at the start of the season would appoint a *shaquq* to head the field. Literally this means the one who cuts, and his duty as the best harvester in the rabta was to organize the movement of the *wajh* (face) or line of harvest so that it remained continuous until the whole field was harvested. In his work he had therefore to take into consideration the contours of the field, the roads and other fields surrounding it, as well as wind direction and sunshine during the different times of the day. At the end of the line was usually placed one of the older harvesters and he was called *al-jahush* or the donkey-rider, meaning that he slowly followed the group.

To be absolutely sure that the harvest was carried out in time, the *mu'allimin* (landlords) decided in conjunction with the shaykh al-rabta, especially in good years, what number of extra harvesters was to be employed. These were paid for by the household and thus the harathin quarter (share of crop) was not affected by them. Generally every rabta had between 6 and 12 extra harvesters. The line of harvest in a big field would have the original 24 harathin together with the extra harvesters in one straight formation. Every harvester had a space of about 3 metres, so the length of the line would be some 100 metres in all. Behind this line there would be a group of 8 *ghamarin* (bundle gatherers), two *shaddadin* (net handlers), one *jammal* (camel man), the shaykh al-rabta and at least one *waqqaf* (overseer) who was normally armed and mounted. Usually it was preferable to have girls to gather the bundles from behind the harvesters as they were more patient in the collection of crops from the ground as well as being better for the *haqla*'s (harvest field) morale. Sometimes they would sing and the men were careful not to show signs of fatigue in front of them.

The operation, although simple, had to be carried on continuously in a systematic manner. The harvesters in the line used sickles. They normally cut the standing crop at the lowest point, because the household needed the largest possible quantity of *tibn* (hay) and *qasal* (straw). They put on a *hawra* (leather apron) to protect their clothes and many of them tried to protect their hands by placing on them primitive home-made leather gloves. On their heads all the men in the fields alway wore the *hatta* and *'iqal* (kufiya and headband) placed in different positions for protection against the sun on these hot summer days. The girls covered their heads with pieces of cloth. All these precautions, however, were of little comfort when work was carried out in the open fields at noon and the dry, scorching heat caused the temperature to rise to 50 degrees centigrade or more. The continuous mobility of the haqla did not permit the erection of any *buyut sha'ar* (black tents), and therefore everybody had to manage as best they could during the lunch recess or afternoon break by finding any possible shade under which to lie down and rest.

The ghamarin followed the *hassadin* (harvesters) at close range. It was considered good performance for the hassad to gather the handfuls of stems, each of which was the result of a sickle-stroke, in a *ghumr* (bundle) of reasonable size. This being the case, the bundle gatherer would have less walking to do and she would pick up every bundle in one movement. She would then move to the net which had already been laid on the ground by the shaddadin. The bundles were placed on the net, which was composed of four wooden, round sticks 1.5 metres long by 6 centimetres diameter bound by ropes in the form of a net that, once filled, would form a rectangle of 1.5 metres by 1 metre. Immediately a net was full, the two shaddadin would wind the ropes around it, each having one foot firmly placed on its side, and pull the ropes until the net was well bound. These full nets lay there until the camels arrived. Two of them were loaded on every camel by the camel-man and the *rajjad* (or *rajud*, camel-driver). The duties of the rajud were to convey the load of his *maqtar* (pl.

maqatir, camel caravan), normally three or four in number, to the baydar. He was usually allocated a donkey to ride and, depending on the distance between the haqla and the baydar, he could make between three and six trips a day. Every rabta had between three and four of these maqtars and the shaykh al-rabta had to arrange in advance with his colleagues the number he needed, depending on the distance between the haqla and the baydar.

During the two long, hard months of harvest every member of the household acted to keep the work in full swing. The *wakil* (steward) had to see to it that the *saqqa* (water carrier) with his two donkeys was there in time and that there were enough containers in the field; these were always moved with the harvest line. The *qatruz* (stable boy) carrying the food was expected to arrive exactly when the work force was ready to eat. A sufficient supply of nets had to be ready on the spot and an extra sickle to replace a broken one must be ready to be dispatched from the village at short notice. First aid facilities were never thought of, and if an accident happened, such as a wound while working, a kick from an animal or a snake bite, the injured had to walk back or be mounted on a donkey and brought to the household where iodine was applied, or a herbal medicine given.

All through the harvest, a swarm of *laqatat* (gleaners) followed the line of harvest at a respectful distance. They could easily number 100 women and children who, after the nets were loaded on camels, moved in to pick the fallen grains or intact stems that had been overlooked by the harvesters. Generally they were families of some of the work force as well as poor people who lived in al-Yaduda during the summer. Normally they appreciated the privilege of being allowed to pick behind the haqla and the clever and hard working among them might end up in good years with 1–2 sa‘s of seeds (6–12 kilograms). This was nearly equal to what an ordinary worker could make in wages for a day's work. Such extra income was greatly needed to improve the family's purchasing power.

Food during harvest time was of exactly the same standard as during the rest of the year, with a little extra variety as a result of the availability of some vegetables from the *maqathi* (summer vegetable fields) that gave a crop of marrows, *faqqus* (a variety of cucumber), onion, garlic, eggplants, beans and sometimes okra. These products were never sold and any surplus quantity after household use was distributed to the different families in al-Yaduda. In addition, many donkey loads of green and black grapes arrived daily from al-Salt and Fuhays and these were bought on a barter basis of so much wheat or barley for two boxes of grapes, each of which would weigh 15 kilograms. Around the 1930s the current rate was one *mudd* (3 sa‘s or 18 kilograms of wheat) per box and as a sa‘ was worth nearly 5 piastres the whole donkey load was worth 30 piastres.

Just before sunset every day – there was no holiday or weekend – the rabta stopped work. The expression used in Arabic is *halat*, meaning unwind, and since a rabta in Arabic means literally a bundle, it will be easily understood how these agricultural expressions came to be generally used. The trip back to the village was made by most of the men and the swarm of laqatat on foot. In some cases the distance was short but from the outlying fields it was sometimes a good hour's walk. Naturally the whole work force was tired and hungry on arrival. The women workers and the laqatat each went to their houses or tents while the men, with very few exceptions, went to the household where they washed, using water which the *qatariz* (stable boys) drew for them from the wells in the courtyard. Within fifteen minutes a hot meal cooked by the *‘izbiya* (cook) was provided. Generally it was *‘aysh* (crushed wheat boiled in water) with a touch of salt. *Qufra* (fat), *samna* (ghee), olive oil or *laban* (yoghurt) was added and the dish was eaten by hand from *ankari* (probably an old Turkish word), copper plates, that could accommodate five or six men each. Sometimes onions were offered on the side. There was always more than the men could eat.

After a hurried meal that was over by 7.30 p.m. the men moved around, and by 8 p.m. many of them would be, after a very tiring day, fast asleep on the stacks of hay at the baydar. Very few of the men slept in the stables during summer and even the stable boys, whose duty it was to be next to the animals, slept in the open courtyard.

The end of harvest came, at long last, during the first half of August. Full preparations were made for the feast called *jawra'a*. A good plot of standing wheat, between 10 and 15 *dunums* (see appendix 2), decided upon by the landlord and the shaykh al-rabta, was declared free for the laqatat. The sign was normally given by a shot fired by the mounted waqqaf. Every year things were planned so that the completion of work took place at around three in the afternoon when the swarm of laqatat, accompanied by another crowd of children and women from the neighbourhood who had heard about the occasion, joined to collect some wheat for themselves on the last day of harvest. The loaded camels, 16 of them, were kept until everybody was ready and then the march proceeded back to the village. On every camel a girl rode between the two nets and she carried a white banner declaring the end of harvest. All the men walked in a formation in front of the camels singing, and the few riders in the caravan galloped a little, every now and then shooting their rifles in the air. This festivity continued until all the rabtas managed to gather in the plain lying to the east of al-Yaduda, only 300 metres away. Riders then came galloping from all round; for the next half an hour a *sabiya* (riding contest) was held when more than fifty riders, including the guests, played around, raced, and performed on horseback. Later the whole group moved in a procession over the short distance to the village where everybody disbanded until the evening. In the meanwhile a she-camel or an ox was killed and the meat cooked in a huge cauldron that was the talk of all the countryside around al-Yaduda. Huge logs of wood were especially used on this occasion for the fire required to cook the food.

The whole village gathered again at around 6.30 p.m. when everybody was offered dinner composed of 'aysh and meat. Normally the meat was boiled in its own fat, but samna and laban were poured over the prepared plates. Once the meal was completed the party started. Usually dancing and merry-making continued until midnight. Two or three days of rest were normally allowed after the end of harvest.

Al-Baydar: hopes, rewards and disappointments

Unlike harvest time, when crop collection took place in many fields, baydar time was spent at a fixed place very near to the village. The word baydar had two meanings: the time when it was carried out and the place where it happened. The site in al-Yaduda, until the division between the brothers, was a large fallow piece of land 100 metres to the east of the buildings. At the turn of the century this became the baydar of Farhan, while Frayh had his baydar to the south and Farah had his to the north.

The west, as in many other villages, was kept free in order to save the village people the unpleasant effects of *ghubra al-baydar* (baydar dust) which otherwise would be blown into the village houses. Westerly winds were common in the area and they were very welcome at baydar time, but if the site was placed on the western side of any village the dust would have inevitably flown eastwards in the direction of the houses. The main baydar, which remained in use until recently, was around 15,000 square metres in area and was completely surrounded by a *sinsla* (stone wall) 1.5 metres high, thus making it impossible for animals to enter except through the single gate which was on the eastern side, opposite the *madafa* (guest-house). The grounds were regularly cleaned of stones, and stone-rollers, originally pieces of Byzantine columns that were lying in the ruins, were used to press the soil every now and then. However, the attempts to make a hard surface of the ground were unsuccessful and every year grass grew in spring, which was cut at an early stage and used as animal fodder. After that the grounds were rolled again a few times.

The position of the different varieties of crops on the baydar was permanent and not changed from year to year. The corner on the north-east was reserved for the *kadis al-qamh* (wheat haystack) which had all the wheat hay in its crude form after the *maqatir* of camels were led in by the rajud and the nets emptied all round. When the heap grew in size, camels were led on top of the kadis and their loads were emptied there. In a good year this pile of hay rose to a height of over three metres. To help the rajjad unload the nets from the camels' backs, one of the three or four assistants of the *bayadiri* (baydar supervisor) went up on the kadis. Once the unloading was completed the empty nets were loaded on one camel and the maqtar departed to the field to bring a new load. The *tarha* (heap) of barley, lentils, chick-peas and vetches which had already been harvested were in the meanwhile being worked upon and some of the seeds had already been stored.

The *diras* (threshing) started practically the same day the harvest was beginning in the field. This was carried out by mules only, whereas in other communities people sometimes used oxen, cows and donkeys. Pairs of mules were bound to a *lawh diras* (threshing board), which was two planks of heavy wood fixed together by nails on the upper side, with sharp pieces of flint placed tightly in holes on the side of the planks resting on the floor. The board destroyed all kernels or straws with which it came in contact as it was drawn by the two mules. A *darras* (thresher) normally stood on the board carrying a *miqra'a* (a crude whip) with which he could hit the mules every now and then to make them move faster. A tarha (hay batch) was normally rounded into a circle when some 500 nets had been emptied on the ground. The threshing continued as the bayadiri and his assistants turned the hay every now and then with their *shawa'ib* (pl. *sha'awb*, an implement made of four prongs of steel and usually imported from Damascus, Nazareth or Nablus). Two or three teams of mules were placed on such a tarha and it was normally ready for *dharra* (sifting) within three or four days. To save time and not allow the mule teams to be without work, they normally were then placed on the kadis and the whole tarha gathered in one straight heap of 20 metres in length by 2 metres in height. With the wind blowing from the west, the bayadiri and his assistants would start the dharra, each using a *midhra* (pitch fork) which was four wooden prongs in the shape of a triangle held together by a piece of raw hide. Work continued in this fashion as long as the wind blew and within two or three more days the line of golden hard wheat would be clean in one long row with parallel lines of qasal and tibn lying to the east of it. The qasal was used in mud bricks and heating the *tabun* (baking oven); the tibn was used solely as fodder for animals.

Once the tarha was sifted, the wheat was cleared by the use of a *ghirbal* (sieve) which allowed the wheat seeds to slip through its openings while the larger stones and straw were held. This agricultural implement, together with the *kirbal*, which had larger openings and was used for the sifting and carrying of straw, was normally made by craftsmen of the gypsy clans[17] who visited al-Yaduda every year and manufactured the required number of implements for a whole season. In addition they produced *manajil* (sickles), *madhari* (prongs), knives, *manaqil* (ember containers) and wooden spoons for cooking called *khashuqa* by the locals.

Work at the baydar continued at a relaxed pace as long as the harvest went on, but the tempo completely changed as soon as the jawra'a and its two or three days vacation were over. The whole basic labour force of 78 men and their *shuyukh al-rabtat* (rabta chiefs or foremen) were now fully engaged on the bayadir. The place became a beehive of activity and work continued late every night. This suited the camels which had to carry the load of seed in *'udul* (sacks) or qasal and tibn in *khiyash* (pl. of *khaysha*, a huge bag made of jute) from the baydar grounds to the stores on the top of the hill. Besides, work on the baydar didn't normally start until the sun had evaporated the *nada* (dew) from the haystacks, and this meant the men could sleep or rest until around nine in the morning.

Once the seeds were sifted and ready, the heap was called *subba*. The produce was immediately loaded into the stores so as to avoid any surprise by a raiding party, although this was not frequent at al-Yaduda. In the twentieth century, however, this procedure was discontinued and a subba was left on the baydar grounds for a day or two depending on the work programme. In that case a system was developed to safeguard against pilferage; a *khatm* or *rashm* (wooden seal) was used to stamp the seeds at a level of 5 centrimetres higher than the ground all round. The seal had on it Koranic *ayat* (phrases) such as 'La Illah ila Allah' (There is no God but Allah), or 'Bism Illah al-Rahman al-Rahim' (In the name of God the gracious and most merciful). Normally the seals were made to order in Damascus. The shuyukh al-rabtat were responsible for seeing that the stamping was carried out at sunset and that the seals were complete and not tampered with on the morning of the following day.

Storage of seeds was in bulk in built-in compartments called *qutu'* in the buildings, or in special wells of which the bottoms were completely covered by qasal before the seeds were heaped in. Qasal and tibn likewise were stored in wells but a good quantity was also placed ready for use in the *bawayik* (sing. *bayika*, stables). Naturally a good part of the crop was sold every year to the visiting beduin tribes who came to Al-Balqa' for the dual purpose of selling some of their herds and buying wheat for their food requirements and barley for their horses. The process of wheat purchase was called *yiktallu* (to buy by the measure) and occasionally it was done on a barter basis. Farmers generally required a small supply of good camels for their work. The guest-house hospitality tradition at al-Yaduda must have been a very important element in this necessary trade, with the produce and the quality of the seeds enhancing the continuous demand for its wheat. The semi-settled tribes and settled population, on the other hand, bought wheat and other seeds regularly.

All these sales that were carried out during the baydar times were measured and accounted for by the wakil of the main household. All quantities entered in the stores or wells were similarly registered. The unit of measure was the sa‘ (6 kilograms of wheat) although the actual measurement by the *kayyal* (he who measures cereals) at the baydar, who was usually a shaykh al-rabta, was by the mudd measure (three sa‘s). The first measure removed from a subba was called *sa‘ al-Khalil* (Abraham's sa‘) and was placed in a bag and given to a religious man, if there was one around, or to one of the crowd of poor visitors received at al-Yaduda every summer. This tradition was thought to provide a good omen and was always adhered to. After that, every 8 mudds were placed in an *‘idal* (a bag made out of woven wool cloth, which was produced in villages and encampments) and two of these containing a *batiha* (48 sa‘s) were carried by a camel to the stores, thus making counting of the measures of produce a simple operation.

Towards the end of the baydar, an estimated quantity of grain equivalent to the quarter owing to the labour force was left heaped on the floor and each shaykh al-rabta as well as the murabi‘iya were issued their respective shares. Generally every one of them reserved two or three batihas (96–144 sa‘s) for his family's *muna* (supply) during the coming year and sold the balance to merchants. These could be from among the farmers or visiting dealers from al-Salt, Amman, Jerusalem or Nablus. Generally everybody heaved a sigh of relief when it was said at the end of September or the beginning of October: *tayarat al-bayadir* (the threshing floors have flown away).

4 Land: Ownership and Acquisition

What is land worth without a government?

The last two decades of the nineteenth century and the beginning of the twentieth witnessed a certain consolidation of the demand for land for agricultural purposes. Evidently the increase in population as a result of natural growth and migrations from the west and the south[1] made people realize, more and more, the inherent value of holdings in the name of the clan or the village.

Conditions in the western parts of the northern area were partly stable: villages and *nawahi* (sing. *nahiya*, sub-district) were already established units under the musha‘ system and ownership was rather well defined among the different families in every village. However, this did not mean that certain acts of acquisition by force did not take place. A good example of this is that of Yusuf al-Barakat, of al-Frayhat clan, shaykh of Jabal ‘Ajlun, whose first act when he assumed leadership was to annex the lands of Kufranja[2] which belonged to the villagers of ‘Anjara. In 1812, Yusuf was the lord of Rabad Castle,[3] just outside the town of ‘Ajlun, and was thus strong enough to force the issue against the weaker inhabitants of these two neighbouring villages.

In the middle area of Transjordan the standard of stability regarding land ownership was much lower. Except in the immediate area around al-Salt and the tribal lands of semi-settled beduins like the Balqawiya, Da‘ja, ‘Ajarma and Ghunaymat, land rights were being contested by the ‘Adwan

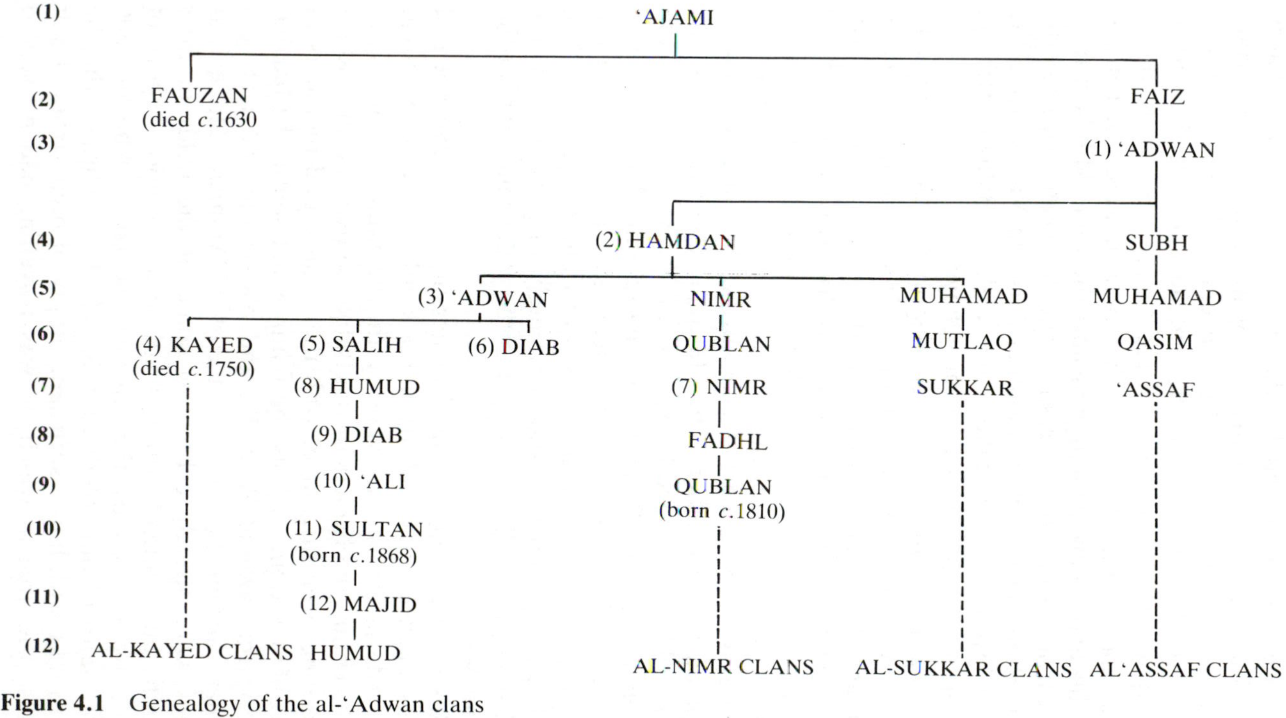

Figure 4.1 Genealogy of the al-'Adwan clans

Note: Numbers beside names denote that shaykh has been shaykh al-Mashayikh (paramount head) of al-'Adwan.

in the western areas and by the Bani Sakhr in the eastern areas. The 'Adwan (see figure 4.1), who became an important local force towards the end of the seventeenth century,[4] drove off the Mihdawi tribe from the plateau and forced them to settle down in the southern Jordan Valley areas of Kafrayn, Rama, and Nimrin. However, the 'Adwan were not content with this and a new campaign against the same Mihdawi tribe was carried out. As a result, the 'Adwan became the uncontested masters of the southern Jordan Valley while the Mihdawi were forced to take refuge with the Balawna tribe in the northern Valley area of Abu 'Ubayda.[5]

The newly acquired lands were distributed among the clans of the 'Adwan some time during this period around 1760. Salih and his younger brothers, the sons of 'Adwan, got Ghawr Nimrin; Kayid, the oldest son of 'Adwan, got Ghawr al-Rama; and their cousin Qublan, the son of Nimr, and his brothers got Ghawr al-Kafrayn. Even then, Kayid was not satisfied as he believed his share to be the smallest, and he complained to his father. The old man could offer only one solution. He encouraged his son to raid the Mihdawi tribe and their new hosts, the Balawna, one more time. The advice was taken and after a few bloody encounters the 'Adwan were able to take hold of all the lands up to the Zarqa' river. The Mihdawi, impoverished and completely dispossessed of any land, migrated to the Bisan area in the Palestinian part of the Jordan Valley.[6]

This drastic change in land ownership must have been of great influence on agriculture in the whole of the Balqa' area. The 'Adwan clans, who were by the beginning of the eighteenth century the leaders of the Balqa' confederation, had for some time been in contact with the semi-settled members of the confederation, such as the Ghunaymat south of Madaba; the 'Ajarma around Wadi Sir and Na'ur; the Da'ja around Amman; and al-Balqawiya just east of Amman. These four groups were already cultivating some of the fields in their domains but evidently on a small scale. A good report of conditions is given by the traveller Seetzen,[7] who in 1806 made the trip from al-Salt to al-Karak. He confirmed that he had to take with him a sufficient supply of bread from al-Salt because the area in which he was travelling had no settled population. He spent two nights with a Bani Sakhr encampment near Wadi Sir, then continued southwards to al-'Al, Hisban, Madaba, Ma'in, Wadi al-Wala, Dhiban, Wadi al-Mujib, an encampment of the Bani Hamida, Shihan, Rabba, and al-Karak.

This first-hand report leads us to conclude that it was during the first half of the nineteenth century that the importance of land ownership began to be realized, after the 'Adwan thrust into the valley. It is indeed probable that in the first place the 'Adwan were interested in asserting their influence and leadership through the removal of a permanent threat and contender for the leadership of the area – the Mihdawi tribe – but the fact remains that, once they had the land, they started making use of it. The different weak and impoverished clans of the Ghawr area, together with the 'Adwan slaves,[8] were the new labour force that started the agricultural operation on a worthwhile scale in the eastern Jordan Valley.

It was probably before 1800 that an important development involving land division among the three most important factions in the Balqa' province took place. The first group, the Saltiya (people of al-Salt), were growing in number and importance, thanks to the newcomers and the population's natural growth. They were already spread in all directions around their city, sometimes to about 30 kilometres distance. They were good, hard-working farmers and sturdy enough to stand firm against semi-settled and nomadic tribes. They also held a strong card in their hand, in that their town was the only market-place in the area and was necessary to both the Balqa' tribes and the Bani Sakhr clans. Actually they were becoming active merchants among the beduin; the number of houses engaged in trade with them around 1812 was already about 20.[9] They also acted as middlemen between the Nablus and Jerusalem districts and the nomads of the eastern areas, especially in the *qilw* (soda ashes) trade; around 3,000 camel-loads were sold to the soap manufacturers in Nablus yearly. About 500 camel-loads of sumach (*summaq*) leaves[10] were also sold by the Saltiya to the tanneries in Jerusalem every season. Raisins processed from grapes grown around al-Salt were also exported in good quantities. Mentioning their influence Burckhardt confirmed that the town was collecting a fee from the Bani Sakhr per load of qilu which was used to support the four guest-houses that mainly extended hospitality to the beduins who visited al-Salt.[11]

The 'Abad tribe, the second group in the division agreement, were living all around al-Salt, but on the outside perimeter, and they too had to coexist with the Saltiya as al-Salt was the only place where they could sell their produce and buy the goods they needed. The third group was the 'Adwan, who, although they were the paramount shaykhs of the Balqa', could only exercise nominal jurisdiction in al-Salt itself.[12]

Tribal distribution of land

It is said that the division of land among the three groups was made more urgent by the pressure that was felt as a result of the incursion of the Bani Sakhr into the area in larger numbers, and for longer durations, than had been known until then. Elderly people mention that the division was one among brothers (meaning equal shares): thus the Saltiya, the 'Adwan and the 'Abad each received *maqsam akhu* (share of a brother), so confirming the equality of status of the three.

Hajj Tawfiq al-Riyalat,[13] who confirmed hearing these stories from his grandfather, also gave an idea of the villages that were allotted to each party as a result of this land settlement. The resultant list was later checked by Sulayman Qamuh,[14] and after a few visits to different parts of the area it was possible to draw up a table (see table 4.1) showing the holdings of each group and the areas of land involved. The villages are enumerated from north to south and the area figures shown opposite each are those listed by the Lands and Survey Department of the Hashemite kingdom of Jordan. It must be stated that the term 'village' is here loosely used to denote an area that only became inhabited towards the end of the nineteenth century. Before then they were ruined and uninhabited sites whose fields were cultivated by farmers who either resided in al-Salt, in the case of al-Saltiya,[15] or in beduin encampments, in the case of the 'Adwan and 'Abad tribes.[16]

Table 4.1 Villages allotted to the three parties in the land settlement in al-Balqa' towards the end of the eighteenth century (areas in dunums)

'Abad share		'Adwan share		Al-Saltiya share	
Village	Area	Village	Area	Village	Area
Subayhi	10,852	Mubis	11,318	al-Rumman	21,495
Bayudha	53,202	Abu Hamid	2,572	al-Mastaba	14,416
Sihan	19,449	Abu Nausayr	17,397	Jal'ad	20,054
Sumiya	8,968	Badran	4,161	Salihi	9,639
Maysara	32,526	Umm Zuwaytina	4,161	'Allan	6,510
Mahis	21,306	al-Hummar	3,567	Umm al-'Amad	4,631
'Ira	24,318	Jubayha	11,775	Rumaymin	7,357
Yarqa	17,599	Yajuz	1,653	Umm Jawza	7,357
Humrit 'Ira		Swaylih	3,743	Da'am	893
and Yarqa	52,503	Khilda	4,438	Umm al-Dananir	10,415
Bilal	46,772	Dabuq	6,141	'Ayn al-Basha	21,414
Al-Bassa	18,309	Tula' al-'Ali	10,097	Al-Salt	210,424
'Iraq al-Amir	34,056	Ghawr Nimrin	78,841	Safut	7,947
Al-Bahhath	34,876	Ghawr al-Rama	63,319	Fuhays	23,580
Wadi Sir	36,556	Ghawr Kafrayn	39,900		
		Hisban	1,440		
		Zabbud	44,681		
		Na'ur (half)	23,222		
Total	411,292		332,426		366,132
Total hectares	41,129.2		33,242.6		36,613.2

Note: 1 dunum = 1,000 sq. m.

With the basic division of land among the three parties completed, the stage was now set for further subdivisions among the different clans within each group. Matters were more or less clear among the three major clans of the 'Adwan and the two large tribal groups of 'Abad, each of which had their domains vaguely defined by tribal tradition. For the people of al-Salt, however, it was necessary to make special arrangements, taking into consideration the continuous changes in the demographic conditions in the town itself and the traditional divisions among its different *hamayil* (clans) and families. The increase in population figures was one of the important elements that influenced the final allotments. Elderly people of al-Salt relate that the dignitaries of the town met and after due deliberation agreed to divide the lands allotted to al-Salt into two halves. One of these was to be the share of al-Akrad groups and the second was to be that of al-

Hara groups, composed of the tribal rally of the 'Awamla, Qutayshat, and Fawa'ir. The most interesting feature of this arrangement was that land was allotted to those clans and families who were farmers. Thus the Haddad clan of the Habala quarter, who were part of al-Akrad group and earned their living as village blacksmiths, were not allotted any agricultural land. On the other hand, another Haddad clan of al-Khadir quarter, whose members had probably ceased being blacksmiths before their arrival in al-Salt from al-Karak a century earlier, and had become part of al-Hara group, were allotted land as farmers.

The distribution of land to individual households was made by the heads of clans only as a third stage. It is said that when the share of al-Akrad half was to be divided among the different hamayil, the shaykhs decided first on the principles for such a division. After some consideration, it was agreed that the half would be divided into ten roughly equal shares. Nine of these were distributed to the nine major tribes: al-'Arabiyat, al-Jazaziya, al-Khuraysat, al-Hayasat, al-Quama-qima, al-Nusur, al-Ramamna, al-Hiyarat, al-Dababsa. The remaining one-tenth was allotted to the four smaller clans of al-Akrad group, who as of that time became known as *Riba' al-Hamayil* (the one-quarter clans).[17]

The same principle must have been in some way or another applied in the case of al-Hara group, although the stories are vague and do not give as clear a picture of what happened with this group. However, it should be mentioned that the Christians of al-Salt, who made up one-fifth of the town, were mainly members of al-Hara group, and their clans[18] received their shares of agricultural land from al-Hara's share. Belonging to one group or another was, in earlier times, dependent on the quarter in which a clan resided. There were only two Christian clans in al-Akrad quarter.[19] These were al-Haddadin, who did not seek a share because of their profession as blacksmiths, and al-Qamaqima, who were one of the nine major clans in the section and received their equal share of one-tenth.

There was in al-Salt generally loyalty to the quarter and its clans which was determined, first and foremost, by the proximity of residence. There was also a strong feeling of mutual interest in security and matters of common involvement among those who belonged to the same quarter, irrespective of their origin or denomination. Therefore, religion played a minimal role in the political affairs of the town and this provided its people with the means and determination to stand united against outsiders, whether the beduin confederations surrounding them or the Ottoman walis and the pressures of their military expeditions.[20]

Acquisitions similar to these are still called 'tribal divisions' nowadays.[21] They must have occurred among the people of Transjordan all through the second half of the eighteenth century and the first half of the nineteenth. These were the days when land, especially in the plateau area extending to the east of Irbid, al-Salt and al-Karak, became important for agricultural cultivation rather than merely providing pastures for flocks of sheep, goats, and camels.

Use of force and litigation

Differences between tribes, clans and individuals arose frequently and were generally settled either by force or by the parties resorting to litigation under the auspices or in the presence of renowned tribal judges.[22] Amongst examples of settlement by force are the two incidents mentioned at the start of this chapter and the takeover of the lands in the northern area of al-Karak district. The latter issue was decided through tribal warfare when the tribesmen of al-Karak defeated the Bani Hamida at the battle of Qadduma some time after 1850.[23] This, as in similar cases, forced the defeated to migrate to their present domain north of al-Mujib river.[24]

An example of settlement by litigation is the dispute that raged between al-'Amr and al-Majali tribes in al-Karak area around 1780. After many encounters the two parties agreed to attend an oath session in which the representative of the claiming party had to define the boundaries and swear that his markings were true. Once in the field, Hamad ibn Majali, on behalf of the Majali, suddenly stopped marking the boundaries and started placing sand in his shoes. He explained that his action was necessary to allay the evil omen caused by a crow that was hovering overhead and croaking. Half an hour later he completed his work and swore that the lands on which he stood belonged to his party from the time the crow croaked. In their innocence, al-'Amr accepted the oath; so Hamad managed to win for al-Karakiya (the people of al-Karak) the vast area south of their town. Afterwards al-'Amr, who were shaykhs of al-Karak, were dispossessed of the remainder of their lands and the remnants of their tribe were forced to migrate to Palestine. Decades later they were invited to come back, but as vassals of al-Karakiya[25] and not as the masters they used to be.

An interesting land dispute that extended over 40 years mixed the two elements of force and litigation. It was of great importance to the two contesting parties, not only because it involved 4,000 dunums of the best wheat lands between Amman and Madaba, but also because it would set a precedent that was expected to apply to most of al-Yaduda's 22,000 dunums. That is probably why most of the case documents are still preserved after nearly 90 years. The case resulted from a purchase of land that was agreed upon between Fawwaz al-Fayiz, of Bani Sakhr, and Farhan Abujaber and his two brothers Frayh and Farah in the year 1319 H/1901 AD. The oldest document in our hands is a copy of the sale deed through which four of Sattam's sons, Jurayid, Shibli, Mithqal and 'Arif, authorized their older brother, Fawwaz, to sell their shares of the lands called Rujm al-'Arayis for an amount of 500 Majidi riyals or 12,000 piastres. Evidently the shares of the three older brothers, Fayiz, Fawwaz and Nawwaf, had been sold earlier, when 'Awad, the youngest brother, was only ten years old. This document, being final and legalized during a *majlis* (council) held at al-Yaduda in the presence of the five Fayiz brothers, was thereafter entered in the registers of the Registry Department at al-Salt. The Abujabers continued to cultivate the land and, following the customs of those days, paid the equivalent of the '*ushr* (one tenth tax, or tithe) to the Fayiz brothers on a yearly basis. As a nomadic tribe, the Fayiz brothers did not pay agricultural taxes and were only held responsible for the animal tax accruing on their flocks and herds; thus they acquired an additional annual income. This system continued for many years and there are many receipts for all the Fayiz brothers including one signed by Mithqal himself, dated 2 August 1326 H/1908 AD.

During 1909 there was a sudden development when Mithqal, gathering a large number of his tribesmen, prevented the Abujaber ploughmen from cultivating the land all through the winter season. The Abujabers brought the matter to the knowledge of the Ottoman authorities in al-Salt. The qa'immaqam issued instructions to Fawwaz ibn Fayiz, Mithqal's brother, who had by then become a *mudir al-nahiya* (governor in the Bani Sakhr area) at al-Jiza. The pressure seems to have succeeded as we find among the documents one signed by Mithqal, on 12 April 1910, admitting that he was the aggressor and undertaking not to obstruct the cultivation of the land any further. However, on 14 April 1910, Fawwaz wrote back to the qa'immaqam complaining that Mithqal was persevering in his course of aggression and strongly recommending that the authorities issue instructions to the responsible officials (meaning the army) so that Mithqal would be given a harsh lesson, especially since he was the aggressor and the land had long ago been legally bought by the Abujabers. He furthermore requested that they send him an additional force of 50 soldiers since the number at his disposal at al-Jiza was definitely not sufficient.

The Ottoman authorities do not seem to have been in a frame of mind to force the issue and therefore friends of the two neighbouring parties attempted to resolve the dispute in an amicable way. In their enthusiasm to arrive at a settlement, they said that Mithqal had not truly sold the land because his share of the price had not been given to him. Another argument was that the land was very valuable and the Abujabers could afford to pay an extra amount to solve the problem. After a year's delay, and the use of many intermediaries, it was agreed that the Abujabers would give Mithqal a flock of 200 ewes. Ever since that deal the land has been known as *al-Nu'ajiya*, or 'the mother of ewes'.

It now seemed as though matters had been well settled regarding this piece of land. In the meantime Fawwaz died, as well as Farhan, Frayh and Farah. In 1921 Mashhur, the son of Fawwaz and the young *shaykh al-mashayikh* (paramount chief) of the Bani Sakhr, was killed by a member of the tribe.[26] Mithqal became shaykh al-mashayikh in the same year, and experienced as he was in power politics, he must have realized the weakness of the government that was being established in Amman and elsewhere under the leadership of British officers. The police force in Jordan, then numbering 100, was commanded by Lt Col. Peake, the author of *Tarikh sharq al-urdunn wa-qaba'iluha*. During these days of turmoil, and immediately after Mithqal became paramount chief of the Bani Sakhr, a revolt broke out in al-Kura district when Kulayb al-Sharayda defeated the small force that was dispatched to subdue the area in May.[27] In August 1921, a few thousand Wahhabi warriors invaded Transjordan, but Mithqal's tribe, together with other Transjordan tribes, fought gallantly against them and succeeded, with British military assistance, in stopping their advance a few kilometres south of Amman.[28]

It was under these circumstances that Mithqal renewed his claim for the land and again prevented its cultivation by force. The government in al-Salt was informed of the act and Peake, with his 100 men,[29] advanced to the site, in response to a governmental command to restore order. Seeing that the Bani Sakhr were already gathered at Umm al-'Amad in large numbers, he decided to adopt a diplomatic approach. Therefore he rode, unarmed, with only four men, to discuss the matter with Mithqal. After some fruitless conversation, Peake found himself a prisoner locked up in the Shaykh's stables for a day. He returned to his force and withdrew to Amman. The case dragged on for another three years and then a stronger government succeeded in settling the affair. Years later, Major C. S. Jarvis, who like Peake was a British officer in the area, wrote about this incident in detail: 'The land case at es-Salt was in due course settled satisfactorily, and at its conclusion Mithqal Pasha was compelled to apologize to Peake publicly for his behaviour'; he does not mention that it required five long years to achieve this result.[30] Even then, the Nu'ajiya case was not over. On 8 May 1939, 'Awad, the youngest brother of Mithqal, brought an action against the Abujabers. He stated that his older brothers had sold his property when he was a child, and since he had not received any money from that sale, he claimed he was entitled to regain ownership of his share, being one-eighth, or 500 dunums. That share was valued then at one Palestinian pound per dunum, or 500 Palestinian pounds, equivalent to 500 pounds sterling. The case, together with many others, was studied by the Land Court and the verdicts were all in favour of those who were cultivating the land. Forty-one years in the case of al-Nu'ajiya was considered enough time bar to confirm the right of ownership.

Similar cases happened often in the Transjordanian countryside, although in none of them were the amounts involved as great. Generally they arose between the semi-settled tribesmen and their descendants on the one side and the newcomers or settlers on the other. Tribal judges sat on these cases well into the twentieth century, when the general trend has been towards compromise and extra payments by the settlers. When the Ottoman authorities gained stronger control of the whole area towards 1900, the courts started dealing with these cases in a more systematic manner. However, the improved state of affairs was disturbed by the first world war and the years of instability that followed it. It is reasonable, therefore, to say that ownership in Jordan became a matter for law and legal procedures only by 1930.

Gifts and dowries

Gifts for religious purposes were not common, probably because the presence of the religious establishments was not strongly felt in the simple daily life of the Transjordanian communities during the nineteenth century. Research in the registers of the Islamic Religious Court in al-Salt for the period 1302 H/1884 AD to 1305 H/1887 AD did not reveal a single gift of this kind. This is a strong indication that endowments bequeathed for purposes sanctioned by religious law (*waqf*) were very few. Christian churches did not have much estate in the three areas of Irbid, al-Salt and al-Karak until the 1960s. The Orthodox patriarchate of Jerusalem had no property in the whole of al-Balqa' other than the old St George Church and Sarah Cemetery[31] in al-Salt and three churches in al-Fuhays, Safut and Madaba.[32] The Latin, Protestant and Roman Catholic churches generally bought the sites on which they built their churches and schools. Similar conditions prevailed in the villages that had Christian communities in the north, such as ‘Ajlun and al-Husn, and in al-Karak, which was the only place that had a Christian community in the south.

Other gifts were made as gestures, such as the grand gift of land by ‘Ali al-Dhiyab of al-‘Adwan to Sattam of the Bani Sakhr, who was also the husband of ‘Ali's sister. There were also smaller areas involved in gifts to the shaykhs of the smaller tribes, and these were generally called *kabra* or *sharha*.[33] They were gifts by one or more people to a shaykh or well-established man, in recognition of his services to the clan. Such a man may have been the one who ran the estate, was leader of his tribe or had his house open as the guest-house of his community. An interesting case which gives original information about this subject was that brought to court by the descendants of Frayh and Farah Abujaber in 1939, against their cousins, the sons of Farhan. They claimed back three fields, having a total area of 1,000 dunums, which had been given by Frayh and Farah at the turn of the century to Farhan as a kabra in recognition of his senior status among them. The main point in their petition to the Land Court was that the intention of the gift when made must have been the right of use; since Farhan had been dead since 1917, the land should be returned to the estate and divided equally among all the inheritors. This was contested by Sa‘id and Sa'd, the two sons of Farhan, and to prove their case of rightful ownership they presented to the court over 40 witnesses from among the shaykhs and leaders of communities in Transjordan. Their testimony, although varying in details, confirmed the popular view that al-kabra was a recognized tradition in the Transjordanian countryside that was permanent and thus involved final transfer of ownership. Many precedents were mentioned where gifts had been made to shaykhs as gestures of recognition. They had all happened prior to the gift to Farhan and had taken place in different areas throughout Transjordan.

After lengthy sittings, the court in its verdict given on 5 December 1939, established that it was convinced, beyond doubt, that the original gift had been free of any conditions and that Frayh and Farah, when giving to their older brother

Farhan, had the intention of doing so as a final and absolute transfer of ownership. It furthermore declared that it was strengthened in this attitude by the fact that Farhan, and later his sons, had paid the 'ushr on the given land for nearly 40 years.[34]

Nowadays, gifts are still made, although on a much smaller scale, but the use of the term *kabra* is rare. The term *sharha* seems to have been dropped completely. This is because it is derived from the word *sharaha*, which means gluttony to the settled population. The word no longer conveys what beduins originally meant by it, which was coveting and desiring to take possession.

Dowries or marriage portions formed part of the recognized procedure of marriage. They were always paid by the prospective groom or his kin to the father of the bride-to-be. In classical Arabic it is called *al-mahr*, but the word more frequently used by the Transjordanians was *siyaq*, which literally meant animals that are driven as a gift. The explanation for this is that in most cases the dowry or marriage portion was settled with goats, sheep and, in certain cases, a mare as well. Among the more established clans who could afford to dispense with land, fields, or sometimes bigger pieces of land, were presented. Nawwaf ibn Sattam al-Fayiz, just before the turn of the century, gave one-third of his holding at Zuwayzia (1,683 dunums) as marriage money for 'Anqa of the Draybi clan of the Bani Sakhr. A short while later, when she died, he gave the second third as marriage money for her younger sister Falha. When he divorced Falha, he asked for the hand of another belle of the Draybi clan and presented the remaining third of Zuwayzia's lands as marriage money for Madhka. Having become completely dispossessed in this khirba, he moved to al-Qastal (29,846 dunums) where he still had a share with some of his brothers and where his descendants are still living.[35]

Another important case, around 30 years earlier, involved the marriage of 'Ayda, the sister of Sattam, to a shaykh of al-'Adwan. Sattam was already married to 'Alya al-'Adwan and therefore the new marriage of 'Ayda was not considered a *badayil* (exchange) for the marriage of 'Alya. As marriage money, therefore, al-'Adwan presented to Sattam's father,

Shaykh Fandi, a good part of Barazayn's fields, a total area of 3,352 dunums.[36] The shaykhs of al-'Adwan kept up this tradition even into the twentieth century. Shaykh Sultan ibn 'Ali al-Dhiyab married a year or two before the start of the first world war and the marriage money he presented was the title-deed to some 3,000 dunums from his holdings in the plateau not far from Amman. Beduins seem to be more open about these matters: whereas the shaykhs of the Bani Sakhr rendered the best account they could, it was not possible to obtain first-hand information about similar cases from other parties, especially the settled population.

Payment of any kind by the bride's family was not known among the Transjordanians and would have been considered a great shame. The general feeling was that good families, after all, do not buy husbands for their daughters. This is why, perhaps, daughters were not generally allowed to inherit land, especially if they had brothers. The Abujabers gave four of their daughters in marriage to the Bisharats between 1880 and 1920. Because they were not allowed to inherit land, the Bisharats, who themselves were also large-scale farmers at Umm al-Kundum, became unhappy and started legal cases. Relations between the two families, in-laws and cousins as they were, were sour and frequently tense. Matters came to a head when the Bisharats brought their claims to the notice of the authorities in 1924. After heated arguments and threats and counter-threats, these differences came to the notice of Amir 'Abd Allah of Transjordan. He kindly took the initiative to convene a meeting which included representatives of both parties, three from each side. After listening to the Bisharat case, which basically requested the transfer of ownership to them of some 10,000 dunums of al-Yaduda's land, the amir asked the Abujabers to explain their point of view. This, it is said, did not take much time. The Abujabers simply asked the amir kindly to enquire of the Bisharats whether they themselves had given land to their own daughters married to other families. On receiving a negative answer, the amir decided that, since traditions were to be respected, the Bisharats should adhere to them without exception. They were not to receive when they were not willing to give.[37] As was the case with practically all the land-owning families in

Jordan, irrespective of whether they were Muslim or Christian, the married daughters of the Abujabers did not inherit any land and had to make do with relatively small amounts of cash. This situation, however, could not continue during the 1940s when the land settlement law was applied. Some married daughters then lodged their claims and inherited their lawful shares.

Land purchases

Since it has been practically impossible to find documents of land transactions before 1880, the study of land purchase activity has not been an easy task. However, it seems certain that the demand for agricultural land to be used as a means of production only grew after the gradual gaining of control of the countryside by the government. This demand was mainly generated by newcomers who were farmers themselves or interested in farming as an economic venture. The semi-settled population had enough free land at their disposal within their tribal domains and the active among them therefore had no need to buy. They could cultivate any extra area that they needed or desired merely by arriving at an arrangement with the other members of the clan.

This abundance of agricultural land is confirmed by interesting pieces which appeared in travel books during the period. One of the travellers, Henry C. Fish, who visited the area around 1870, was struck by the abundance of rich land in the Hisban area, 20 kilometres south of Amman. He wrote:

> The region from Amman to Hisban was full of flowers and vegetation showing the natural fertility of the soil, and we saw four cities (in ruins) which with the four or five thousand acres of rich land where they lie, are offered for sale (by the pasha-governor of Damascus) for four thousand dollars. But what is land worth without a government?[38]

Another important remark is made by Laurence Oliphant who visited the area during 1879. He was impressed by the venturesome spirit of Salih Abujaber and wrote about land and agriculture in the following terms: 'He had no title deeds or other proof of legal possession but seems to take as much land as he likes, securing himself from aggression from the Arabs [meaning beduins] by payment of a certain proportion of his crops, they acting the part of landlord.'[39] Since Oliphant was an advocate of Jewish settlement in Transjordan, it will be readily understood why he did not consider the beduins as actual landlords. He preferred the land to be free from owners.

During this period, land-purchase transactions were concluded either by a special *sanad* (sale document) signed or stamped by the seal or seals of the selling party in front of a few witnesses, or by having the transaction registered in detail at *al-Mahkama al-Shar'iya* (Islamic law court). Although not as binding as present-day laws and land-registration procedures, the two systems seem to have worked well in the society at that time. However, money-lenders, who seem to have been active then, succeeded even through these procedures in acquiring land as well as realizing high margins of profit. One merchant of Jerusalem dealing actively in Transjordan was Hanna Batatu, who acquired 17,000 dunums at al-Tunayb from Shaykh Rumayh ibn Fayiz, better known as Abu Junayb.[40] The story was told of how amounts of gold borrowed by Rumayh were measured through filling the *tarbush* (red fez used during Ottoman times by city dwellers) of Hanna with gold coins. Rumayh surrendered the land when he was unable to pay back through filling the tarbush the required number of times at the date when the settlement of the debt became due.[41] Another important case of similar nature occurred when the excellent lands at al-Mukhayba, in the north Jordan Valley, measuring 50 faddans passed into the possession of Acre merchants as a result of the village population of 150 being unable to pay back a debt of 27,000 piastres.[42] This farm was later bought by two Lebanese families and was only taken possession of by a Transjordanian in 1945.[43]

Mortgages, however, did not always result in a take-over. For six years, Salih Abujaber lent money to the Harafish clan of the 'Ajarma tribe in the hope of acquiring Khirba al-'Umayri which was in the midst of his lands. In 1314 H/1896 AD, when he thought the deal was near completion, a

Circassian officer, Mustafa Agha, who was serving with the Circassian cavalry in al-Karak, paid the Harafish a certain amount and also undertook to settle the money owing to Salih on maturity. Probably it was around this time that the *tapu* (land registry) department at al-Salt became more recognized by those who were buying lands all over Transjordan. Irbid and al-Balqa' had been under government administration since 1851 and 1867 and had operational tapu departments years before al-Karak, which had its first proper administration in 1894 under the mutasarrif Husayn Hilmi Pasha.[44]

The collection of documents at al-Yaduda for transactions made between 1890 and 1910 provides original information about the different types of land acquisition during that period. Generally the relationship between the buyers and sellers started with a loan, such as the one extended to a group of 13 tribesmen of al-Zufifa tribe at Khirayba al-Suq. It involved the delivery of 400 kayls of clean wheat (weighing around 29 tons), which by contemporary standards was quite a large quantity. It was dated 30 September 1901 and the debt was to be settled ten months later. The borrowers undertook to be jointly and severally responsible for the settlement, and the conditions stipulated in this regard were quite binding: 'At the date of maturity, we shall pay the amount, guaranteeing each other, and he who is present (at maturity date) shall settle instead of the others. The owner of the loan shall furthermore have the right to seek settlement in full from whomsoever he chooses.' These 13 tribesmen were all partners in a musha' division of their village and, judging by the time of the year at which the loan was contracted, it is evident that the wheat was needed mainly for broadcasting during the coming season. Six months later, a group of 38 members of the same tribe sold a tribal lot of land for 400 Majidi riyals. Although the area they sold was 919 dunums, a substantial piece of land, it must have been difficult to arrange farming operations among 38 owners who did not have the extra capital required for it in the first place. Every one of the original 13 borrowers is among the 38 sellers; obviously they wanted to sell a piece of property in the hope of maintaining ownership of a larger one.

Another document is dated 1320 H/1902 AD. It involves the

sale by Hindawi, son of Qasim al-Hajjiri, of all that belonged to him, through inheritance from father and grandfather, in cultivable land, caves, built area of village and water wells, for a total of 85.5 Majidi riyals. In other words, the buyer was actually buying a share in the whole property of a village. The document describes the ownership organization in the village by stating that it was basically divided into three. The third of which the sold property composed a part was 12 faddans and Hindawi was one of 12 partners each owning one faddan. On the basis of what was paid for one faddan, the value of the whole village can be estimated as 770 gold pounds, or a rate of nearly 10 piastres per dunum.

A few years earlier than these transactions, a document signed in 1313 H/1895 AD gives the impression that tribesmen were less inclined towards outright sale of their property. Evidently they were in need of capital and seeds to cultivate their fields and were willing, therefore, to make conditional sales or mortgages with the right of use attached to them. The stipulation in the document specifies that the sellers, a group of 24 Zufifa tribesmen, by receiving 25 Majidi riyals and 720 sa'ʿs of wheat (4,320 kilograms) will surrender the right of use of a plot of land to the buyers for seven years. If, at the expiry date, the tribesmen return to the buyer the amounts involved they will regain their land. As in most of these cases, the tribesmen did not return the amounts involved. This happened sometimes because of inability to do so, but generally because of lack of real interest. The 24 tribesmen must have felt the burden of a musha' system – they were probably not greatly concerned to retrieve a plot of land to be shared among 24 partners and felt more at ease by selling. As in many other cases, the land was finally registered as the property of the buyers.

5 Labour and Partnerships

In Transjordan during the nineteenth century agriculture was the main field of human activity, and the greater part of the population was engaged in it. It was run by small farmers cultivating their own shares of musha‘ land in their villages or in the neighbouring *khirab* (deserted villages). In general these were one- or two-faddan operations worked by the head of the family with the help of his brothers and sons. The general rule was one of co-operation within the community and farmers helped each other during periods of intensive work such as ploughing or harvest on a reciprocal basis. Whenever extra labour was needed it was normally hired for the day or season and wages were paid in kind, usually wheat. There was some forced labour among a few of the tribes, especially the ‘Adwan and the Bani Sakhr; they employed their slaves (all black and usually descendants of Africans bought from slave traders) in the cultivation of fields of the ‘Adwan in the Jordan Valley and those of the Bani Sakhr around Madaba. The small numbers of people living in the countryside, together with their entrenched tribal customs, did not permit in any form the development of a system in which the cultivators were treated as serfs.[1]

The agricultural system developed as conditions grew more stable and with the arrival of new settlers seeking farmlands or employment. It was thus towards the 1860s that larger than usual farming operations started to grow in the eastern districts, especially in the area east of Irbid and al-Husn and the area between Amman and Madaba. The owners or farmers of these new estates needed larger numbers of farm-hands to cultivate their fields and to bring in the crops. Naturally they resorted to the traditional system that was common in the settled areas around them to the north and west as well as in Palestine and the Hawran. The system was based on crop-sharing in the form of yearly agreements between the landlords and the people who were ready to do the work. The former were called *shuyukh* (sing,. *shaykh*) if they were beduins, otherwise *mu'allimin* (sing. *mu'allim*) literally meaning teacher, but actually master. The workers were called *harathin* (ploughmen) if they provided only their labour as ploughmen and harvesters, and *shaddadin*[2] or *fallahin* (farmers) if they undertook to provide the resources and do the work required whether by themselves, with members of their families, or with hired harathin. These two kinds of share-cropping each had a name, the first being called *muraba'a* (division into four)[3] and the second *sharaka* (partnership) or *muzara'a* (farming arrangement).[4] Until the start of the twentieth century these two types of arrangement allowed for only minor variations in the general conditions, especially concerning what we now call fringe benefits.

There was no escape from the need to hire extra hands every now and then. Some of their wages were borne by the venture, while the rest were borne by the landlords alone. The tendency prior to 1914 seems to have been simply to earmark a reasonable quantity of the crop for *tahna* (wheat milled for flour) and *masarif filaha* (farming expenses). In 1911, the total wheat crop of a rabta at al-Yaduda netted 3,191 kayls (230 tons): 191 kayls, or 6 per cent, was earmarked for expenses, and labour's share at 20 per cent amounted to 600 kayls.

Muraba'a

A harath in the muraba'a arrangement always committed himself to serve all through the season from the start of ploughing (October) until the whole quantity of produce was removed from the threshing floor (July or August). Since there was one harath for every faddan, each harath was entitled to a portion of the produce equivalent to the result of dividing a quarter of it (*rub'*) by the number of faddans (pairs

Table 5.1 Calculation of murabi''s share: 1900, 1903, 1911

	1900	1903	1911
Produce, gross	100	100	100
'Ushr (tithe)	(12.5)	(12.5)	(12.5)
Deductions for muna and expenses	—	(4.0)	(4.0)
Total	87.5	83.5	83.5
Murabi' 's share	21.9 (25%)	20.9 (25%)	16.7 (20%)
Landlord's share	65.6 (75%)	62.6 (75%)	66.8 (80%)
		+ 4.0	+ 4.0
		66.6	70.8

Note: The landlord provided all capital and expenses, with the exception of the small contribution made by deduction from the murabi''s
share from 1903. He also retained 12.5% to pay the 'ushr. The murabi' received his quarter or fifth and his muna.

of oxen) employed in the cultivation of the land. In other words he received one-quarter if there was only one faddan or one of twenty-four shares if there were six faddans.[5] In addition, he was entitled to free food and lodging, or he could choose to live with his or another family and receive a quantity of household provisions called *muna*. This varied from place to place but seems to have had a more or less fixed quantity of the basic item, wheat, which was usually 8 sa's (nearly 48 kilograms).[6] However, towards the start of the twentieth century this quantity of wheat given out in the form of muna was being deducted from the gross total of the produce; thus the labour force was actually contributing to one-quarter of the cost of the muna.

The fringe benefits and the rub' (quarter) of the murabi'iya decreased during the nineteenth century. In 1810 Burckhardt[7] reported that a labourer (*murabi'*) in the Hawran received at the time of sowing one *ghrara* (nearly 750 kilograms) of corn and after harvest took one-third of the produce, or among the Druzes only a quarter.[8] The master (landlord) paid to the government the tax called *miri* and the labourer paid 10 piastres annually.[9] At al-Yaduda, the records until 1903 account for the rub' on the basis of one-quarter after deducting only the 12.5 per cent of 'ushr tax from the gross

Table 5.2 Share per harath in various rabtas at al-Yaduda, in kilograms, with approximate values in piastres

Year	Estimated rainfall at Madaba Mm	Wheat		Barley and other crops		Total		Rabta
		Kg	Piastres	Kg	Piastres	Kg	Piastres	
1900–01	233	770	735	732	308	1502	1,043	
1907–08	425	1,980	1,897	735	337	2715	2,234	
1908–09	411	1,848	1,771	670	358 + 71 for sesame		2,200	Rabta of 18 faddans
1908–09	411						1,697	Another rabta. No detail
1909–10	367	1,611	1,746				1,900	Rabta of 19 faddans
1909–10	367	1,560	1,440	696	348	2256	1,788	Another rabta
1910–11	525	1,620	1,553		144		1,697	Rabta of 25 faddans
1911–12	385	1,193	1,029		156		1,215	Rabta of 29 faddans
1911–12	385	1,615	1,552	288	144	1903	1,696	Another rabta
1912–13	369		1,211		340		1,551	Rabta of 15 faddans
1912–13	369	1,257	1,202		322		1,524	Rabta of 15 faddans
1913–14	295	893	893		602		1,495	Rabta of 15 faddans
1913–14	295	831	831		612		1,443	Rabta of 15 faddans

Note: The years listed in the left-hand column are rainfall years; the harvest years are the second of the two in each case.

figure of produce.[10] After that date the rub' decreased as an allowance, for the muna was also deducted from the gross produce. This was at the rate of 48 sa's per faddan, or nearly 4 per cent.[11] After 1911 takings decreased again and the share of labour was accounted for at the rate of a *khums* (one-fifth). As of that date the word rub' was often replaced by the word khums, although a harath continued to be called murabi'. This continuous decrease, as shown in table 5.1, was brought about by the rules of supply and demand. In the Hawran that Burckhardt visited in 1810, a man with a sufficient quantity of seeds and a pair of oxen could, security permitting, probably cultivate the area of land he needed without any hindrance. The same applied in that area even towards 1875, [12] but in Transjordan at the turn of the century the ownership of land had become so defined that every farmer had to reckon with the rent to be paid to thc landlord in the form of a share of the produce. Not a single reference has been found where rent was paid in money.

The actual value of the rub' varied from year to year depending on the yield, which in turn depended on climatic conditions. In years of good rainfall the yield could be twenty or more to the measure of seeds, while in bad years of drought the yield might not exceed five to the measure. To obtain the exact figures, a study has been made of the records at al-Yaduda for thirteen years at the start of the twentieth century[13] and the result appears in table 5.2.

The rainfall figures[14] clearly affect the yield during this period which, with the exception of 1900, must have been rewarding. Prices, on the other hand, seem to have been stable during these years, with a small drop during the year of best yield. However, a comparison of the value of money is the only means for us to judge whether the murabi'iya of those days were being fairly remunerated for their work. The register of the year 1910 mentions the wages that were paid to hands engaged in the harvest during that season. Evidently they varied in relation to the value of the crop being harvested, the nature of the worker (whether a man or a woman) and, most importantly, the work itself. A man engaged in wheat harvest was paid around 8 piastres a day, while the same man harvesting barley was paid around 5

Table 5.3 Additional labour in a farming operation

Job name		Sex	Duration	Wages (piastres)	Fringe benefits
Wakil	steward	man	one full year	1,800 per year	extra wheat and lentils
Harrath	ploughman	man	daily work	5 per day	—
Bawwab	doorkeeper	man	one full year	1,200 per year	tobacco supply
Natur	watchman	man	2–3 months	300 for season	—
Najjar	carpenter	man	one agricultural year	2,400 per year	extra lentils
Jammal	camel-man	man	one full year	2,400 per year	extra lentils
'Izbiya	cook-baker	woman	one agricultural year	900 per year	extra wheat
Qatruz	stable-boy	boy	one agricultural year	650 per year	—
Hassad	harvester	man	for harvest of wheat	8–10 daily or 600 for season	—
Waqqaf	overseer	man	for full harvest	300 for season	—
Ghammar	bundle-gatherer	woman	for full harvest	425 for season	—
Shaddad	net-handler	man	for full harvest	425 for season	—
Saqqa	water-carrier	boy	for full harvest	200 for season	—
Rajud	camel-driver	boy	for full harvest	170 for season	—
Bayadiri	threshing supervisor	man	for baydar	400 for season	—
Darras	threshing	boy	for baydar	250 for season	—

Note: To facilitate comparisons, wages, normally paid in wheat, have been entered in piastres at market prices.
Source: Main accounts book of al-Yaduda for the years 1911–12.

piastres a day. A woman harvesting beans and vetches was paid only 3 piastres. A murabi' who averaged 2,000 piastres a year for around 300 days of guaranteed work was receiving well within the average daily wage (8+5+3 divided by 3 = 5.33 piastres). But the most important element of all for a murabi' was security. He was certain of receiving his monthly muna (provisions) as well as a share of the crop at the end of the season. This share may not have been good in bad years, but it seems to have still been better than the risk of having no work for days at a time.

The primitive economy that prevailed in Transjordan towards the start of the twentieth century was mainly dependent on the small family farming units of one or two faddans each. The mainstay of these were the harathin and hassadin who were young, sturdy and hard-working men. The other members of the family, the weaker or older men, women, boys and girls, all helped in some way or another. There was thus no need for any extra hands in these agricultural operations except during harvest time, when in an attempt to bring the crop in quickly, workers from outside the family were employed for a few days. However, this was not the situation when a successful farmer or a landlord was running an operation of a few faddans or more. One murabi' per faddan was needed but additional labour was also required to service the operation, especially if the members of the family were not available or capable of rendering extra service. Additional labour was involved in more than a dozen different jobs, as will be seen in table 5.3. This also provides a description of the work and the wages collected per day, year, or season.

The determination of yearly wages and salaries was naturally a matter of agreement between the two parties within the traditional terms of the muraba'a system. As to seasonal jobs, when the wages were paid for working during a certain season, these were considered within the framework of employment conditions for that season. Most of the time the worker joined the work on the clear understanding that his wages would be *hasab tali' al-si'r*, meaning that they would be determined by the upcoming market price for a job such as he was doing. In the absence of an organized labour market, this was a practical system used to avoid bargaining in every case. Thus if a certain farm needed six harvesters, they might start work by the beginning of June without knowing what their wages would be until 20 June when it became known that the wages of a harvester for that season would amount to 100 sa's of wheat, or nearly 600 piastres. Then the rate would apply to every one of the six and wages would be paid at the end of the harvest accordingly. The determination of what was paid for every seasonal job was usually the privilege of the largest estate in a certain area, whose owners, after consideration of the labour situation, would make offers for the different jobs. Once these were agreed to by the workers and the word went around, employers and workers alike seemed to adhere to the set rates. Disputes and differences of interpretation were generally referred to the owners of the large estate whose word was generally accepted.[15]

The muna was a basic element in any muraba'a agreement because it gave the harath the assurance that he and his family would at least not go hungry in a bad year. Like the rub' this benefit seems also to have decreased. The first clear mention in the books of al-Yaduda is in 1910 when every harath in the rabta run by Farah Abujaber received the following provisions per year:

Wheat	80 sa's	(480 kg)
Millet	48 sa's	(264 kg)
Raisins	6 rutls	(18 kg)
Olive oil	2.5 rutls	(7.5 litres)
Molasses	2.5 rutls	(7.5 kg)
Onions	3 rutls	(9 kg)
Salt	1.5 sa's	(9 kg)

The same quantities were distributed by his older brother Farhan to his men. During 1914, probably due to the first world war, these rations were drastically cut. They were then distributed at the rate of:

Wheat	7 sa's monthly	(42 kg)
Lentils	1 sa' yearly	(6 kg)
Salt	0.5 sa' yearly	(3 kg)

The cut in wheat did not continue for long after the war, and rations of 8 sa's monthly, with an assortment of the other six items, were reinstated in the early 1920s. The substantial ration of 264 kilograms of millet was never again reinstated, however, as the production of this variety was discontinued.[16]

The fallahin

The fallahin were more involved in the actual running of the agricultural operation as they were partners of the landlord and not merely share receivers as were the murabi'iya or harathin. Their social status was higher and they had the right

بتاريخ ادناه قد تشاركت انا جبر العقرباني وفرحان الوجابرة

واخوانه على ذلك جبر المذكور باراضي السن وصار فرحان ليقدم البذار

الغذا عشر اكيال حنطه وبها قسم ابا بعير نقابه مقسوم بيما بينا

السن ما عدا الحمى الى البخيت وعند جبر المذكور خمسة اكيال

حنطه قرضة حسنة لمضي سبعة اشهر [illegible] تاريخه وليس

على فرحان المذكور لا ربوا ما يعني ولا خليفه ولا يحق [illegible] من الشركا

وهكذا احصل الرضا والتراضي عن بينة وشهود الحال والرحمن [illegible]

٢٢ ايلول [illegible]

القابل بما فيه

جبر [illegible]

شهد بذلك مصلح الزريقات

شهد بذلك فلاح الجبور

شهد بذلك مفلح [illegible]

شهد بذلك جبر ابو زيد [illegible]

شهد بذلك خلف [illegible]

Partnership undertaking

to decide such matters as the timing of the work, the type of crop to be produced, and the labourers to be employed, after consultation with the landlord. Since a fallah ran a much higher risk than a harath, who only provided his own labour, he was naturally entitled to a higher percentage of the crop. This percentage varied depending on the nature of the land to be cultivated, the different items to be provided by each party and the nature of the relationship existing between the landlord and the fallah.

The oldest and only undertaking found among the al-Yaduda documents is shown here. It was signed on 22 Aylul 1315 H/22 September 1897 between Farhan Abujaber and brothers as first party and Jabr al-'Aqrabani, a Palestinian from the village of 'Aqraba, near Nablus, as the fallah. The land was not at al-Yaduda but at al-Lubban, the neighbouring village to the east, owned until the present by al-Bakhit clan of the Bani Sakhr. Under the circumstances al-khums, or one-fifth, of the crop was designated by the two parties to the landlords.[17] The Abujabers were to provide 10 kayls of wheat (720 kilograms) for seeds, and 5 kayls (360 kilograms) as a loan to Jabr, which was to be repaid seven months later, in April 1898. Jabr was to bear all the expenses and the crop was to be split in equal shares between the two parties. Supposing the yield that year amounted to twenty measures, the crop would have amounted to 14,400 kilograms in total, of which the share of the landlord would have been 2,880 kilograms. The balance, divided between the two parties would have given each 5,760 kilograms. For the fallah this was much better than the 2,160 kilograms that a harath would have received for his rub' and muna. The Abujabers, in return for 720 kilograms in seeds and the support given to the fallah among al-Bakhit, received 5,760 kilograms. The agreement does not mention the 'ushr since the Bani Sakhr lands were not yet subject to any taxes.

That this undertaking was put down in writing is due to the fact that it involved also a third party, the landlords. All the other partnerships were oral and the only way to find out about them is through reference to the different entries in the books of accounts. Generally they followed the pattern outlined below, until perhaps the start of the first world war.

Afterwards, new variations started to be applied.

In these arrangements or agreements, land was provided by the landlord or mu'allim, labour, animals, and seeds were provided by the fallah. The crop was allotted to the different parties in accordance with the following account:

	%
Total crop	100
'Ushr al-miri – government tax	12.5
	87.5
Rub' al-harathin 25%	21.9
Balance	65.6

The balance was equally divided between the two parties and thus the landlord's net share was 32.8 per cent. The fallah, on the other hand, received the other half of the proceeds, or 32.8 per cent plus the rub' of 21.9 per cent if he, together with members of the family, were servicing the operation with their work.

The fact that the landlord's share of the benefits of such sharakat was 32.8 per cent (i.e. roughly a third) may explain the tradition prevailing in Transjordan until the present. It is generally accepted that the factors of agricultural production were only three: land, labour and working capital. The last involved seeds as well as animals in general, including those engaged in the two major operations of ploughing and transport of harvested hay and crop. Water was not a factor since all these sharakat were concerned with dry-farming operations, but it was generally understood that drinking water for humans and animals was an integral part of the land.[18] Normally it was drawn from one well or more owned by the landlord on the land or in its vicinity.

Those who provided the three factors were each entitled to one-third of the crop.[19] The only matter to be taken into account was the miri tax, or 'ushr, which in established and fertile areas was deducted from the total crop while in other areas it was wholly borne by the landlord. This was determined by the general conditions in each district and whether farmers were available to enter into crop-sharing agreements. Here again the rules of supply and demand played an important role in creating the conditions of the partnership. Landlords had to remember all the time that their fallahin partners were non-residents who were not forced to stay in any one place. If their remuneration was not up to their requirements and expectations they were at liberty to go to another farm or to return to their villages.[20] It should also be noted that the system was developed after the 1870s when the eastern districts became available for cultivation and that since the landlords were generally tribal shaykhs, the 'ushr, which they did not have to pay, did not become an issue until the turn of the century. Thus the Transjordanian system was much simpler than that which developed in certain parts of Palestine.[21]

It is very difficult to make a guess about the percentage of land that was cultivated on a partnership basis in Transjordan towards the end of the nineteenth century.[22] This will require, due to the lack of records, a detailed study of every village and the details of land ownership at that time. However, it is certain that this form of agricultural arrangement was most common in the area between Amman and Madaba where the shaykhs of the Bani Sakhr did not have the desire, capital or ability to farm the lands that they acquired after the 1870s. It occurred on a much smaller scale in the already settled areas such as Jabal 'Ajlun, Irbid, al-Salt and the western Balqa', where there were only small numbers of large-scale landowners who needed such an arrangement. A simple and practical system, involving agreements that were one-year contracts, it will be readily understood why it has continued until today in the same way.[23]

Places of origin

The frontier of settlement, stretching gradually eastwards from Irbid, al-Salt and Hisban, was being settled as security improved and the growing demand for cereals made farming in the new areas a profitable venture. It is true that some farmers from the already settled western areas of Transjordan moved to the east, but the bulk of the newcomers were from areas across the Jordan River and the Dead Sea. Generally they were from the Palestinian villages, but there were among them farmers who came from the Hawran and Egyptians who, after staying in Palestine for three or four decades, decided to choose a new and more promising life in a virgin land. The Circassians and the Chichen, who were finding it more and more difficult to establish their new homes anywhere, were also among the newcomers. In the two chapters in part three about the Circassians, the Chichen and the Egyptians in Sahab, detailed information is given about their origins, arrival dates and development in the new lands. It is therefore important at this stage to give some information about the other groups and where they came from.

They were mainly Palestinians who came from different villages; the largest number by far were those who belonged originally to the villages of Jabal Nablus. Two of these villages, 'Awarta and 'Aqraba, contributed the largest number of men and it is possible that each had over a hundred men farming in Transjordan towards 1900.[24] Many of them brought their families and thus added to the work force as women and children also worked on farms, especially in spring and summer. There were also villagers from Jabal al-Quds (Jerusalem area) and the Ramallah district. Most people were called after their villages, but men who came from Jabal al-Khalil were known as al-Qaysi, since this area was known for many centuries as a stronghold of the Qaysi faction in Bilad al-Sham.[25] Egyptians were among the first settlers and continued to come as years went by. They were generally called al-Masri, and in a few cases al-Sa'idi if they came more directly from al-Sa'id area. The same general rule applied to those who came from Hawran; they were called al-Hawrani whether they came from near al-Ramtha or from the northern part of Hawran nearer Damascus. Harathin and fallahin who came from Transjordanian villages or encampments were also called after their villages or tribes. For villagers, the most common names were al-Sufani, al-'Ibini, al-'Ajluni and al-Karaki, while tribesmen, who were few, were called al-Sharari, al-Ta'mari, al-Tihawi, and al-'Isa.[26]

The following list of village names in the districts of Nablus, Jerusalem and Ramallah was drawn from the records at al-Yaduda which cover the period towards the beginning of the twentieth century. It gives an idea of the distribution of the most important places of origin of the labour force.

Villages of labour force

Jabal Nablus

'Awarta
'Aqraba
'Anabta
Bayta
'Asira
Huwara
Sarra
Salfit
Balata
Bayt Furik
Qusra
Jalud
Burin
Qaryut
Yasuf
al-Majdal

Jabal al-Quds

Abu Dis
Bayt Iksa
Bayt 'Inan
Bir Nabala

Ramallah

Sinjil
Turmus'aya
Dayr Jarir
Bayt Illu
'Ibwayn
Mazari' al-Nubani
al-Mazra'a

6 Transport and Marketing of Agricultural Produce

The Jordan River and the Dead Sea, which form a natural boundary between Palestine and Transjordan, have been over the ages a barrier against the free movement of people and goods. Crossing from one area to the other in certain districts meant a strenuous trip that entailed descending from an altitude of 1,000 metres above sea-level to points that were 200, 300 or 400 metres below sea-level,[1] and ascending again the same height on the opposite side. The trip required one or two days when travellers crossed between Irbid and al-Salt on the eastern side to Nazareth, Nablus or Jerusalem on the western side. Travel from al-Karak and the southern areas to Hebron and Jerusalem had to be made either around the southern end of the Dead Sea or via the crossing north of it opposite Jericho. Any of these trips required three or four days as the length of the Dead Sea, which did not lend itself to any form of sea travel, was over 80 kilometres. Travel from al-Karak to Gaza across the Wadi ‘Araba also required three or four days.

There were no roads, which meant that there was no carriage transport whatsoever. People had to travel either on foot or on the backs of animals, mainly horses, mules and donkeys in the northern areas and camels in the southern areas. This situation persisted throughout almost the whole of the nineteenth century. Reports of travellers give a clear picture of the difficulties they had to endure in the different parts of Transjordan. Lynch, who went to al-Karak in 1848, wrote: ‘The wheel road means, in that country (Moab), a mule track. Wheel carriages had never crossed it before.'[2] Mary Rogers, a few years later, wrote: 'when I arrived in Palestine in 1855 there was not a wheeled carriage of any kind in the country, not even a wheel barrow.'[3] Although the Circassian immigrants who settled in Transjordan after 1878 brought with them some carriages and a few years later started producing them locally, the traditional modes of travel prevailed. George Adam Smith was induced to write as late as the turn of the century:

> The decay of all these great roads and the disappearance of wheeled vehicles from the land till very recently, was due of course to the conquest of Syria by nomad and desert tribes whose only means of locomotion were animals. The few roads and carriages now in existence are entirely of Frank and Circassian origin.[4]

There were more difficulties apart from the lack of roads and carriage transport; even by 1880, the only crossing point over the Jordan River that had a functional bridge was Jisr al-Majami', just south of the Yarmuk river.[5] The bridge at Damiya, which was originally built by the Mamluk Sultan Baybars in 665 A/1266 AD, had been in ruins for many years and travellers crossing between Nablus and al-Salt used the ford or the ferry.[6] The crossing opposite Jericho used by travellers between Jerusalem, al-Salt and al-Karak was serviced by a ferry boat and a shop. The men who managed the boat also had a hut erected for them nearby and these modern developments led a traveller of the day to write: 'The existence here of even this dilapidated ferry indicates a great advance in the line of improvement for this region. These peaceable boatmen take the place of the robbers who formerly infested the banks of the Jordan and their presence as well as their occupation furnishes our best guarantee of safety.'[7]

These ferries, which were commissioned by the authorities as part of their plan to put al-Balqa' district under their direct control in 1867, were often damaged and put out of use by the rising river. An early report appears in the notes of Lieutenant Charles Warren about the trip he made along the

Jordan during February 1868.[8] He wrote that the ferry at Damiya had been established six months earlier and that in shape it was like a decked launch. Although twenty feet long and eight feet wide, it was eminently unfitted for the work required of it. A thick rope was fastened across the river and the ferry was hauled across by lugging the rope. To the dismay of Warren and his party, the rope was broken on 27 February by the wood brought down by the flood. They therefore had to go north all the way to the Majami' bridge which they crossed on 3 March after five days and many difficulties.

Conditions do not seem to have improved twenty years later. This time, however, it was the ferry in the neighbourhood of Jericho that broke down.[9] The Reverend James Neil was therefore: 'compelled to ride some 18 miles up the west of the Jordan Valley to the ferry opposite *Tell Damieh* and the same distance down the east to the Valley of Ali Diab's camp [Shaykh of al-'Adwan], then in the Plain of Shittim, near *Tell Keferein*. Later on we had to return the same way.'[10] Travel in those days was indeed an adventure involving much hardship and time.

Slow and undetermined as it was, the Ottoman administration took its time to build the first bridge over the Jordan in modern times. It was built, most probably around 1885, to the north of al-Maghtas, a pilgrim bathing place on the Jordan. It was a wooden structure fastened by a gate at each end to facilitate the collection of tolls; the gatekeeper lived in a little hut on the west side.[11] In 1891 the winter rains were very heavy and the bridge was destroyed and washed down to the Dead Sea. For a long time after that the river could not be crossed except by the old bridge below the Sea of Galilee, Jisr al-Majami'.[12] A new bridge was built on a higher level than the old one; access at both ends could be blocked by doors which were opened, when necessary, by the watchmen and tax-gatherer. Although greatly damaged by the swollen river in the winter of 1986–7, it was mended and restored. Considering that the Jordan, in accordance with varying reports between 1867 and 1900, was between 35 and 60 feet in width and around 10 feet in depth, one is surprised by the lack of efficiency on the part of the authorities. The highest authority in Istanbul, the Grand Vizier, addressed a memorandum to wilaya Suriya on 5 Jumada al-Awal 1284 H/1867 AD giving instructions on how to strengthen security and build a bridge over the Jordan. This was to be financed by borrowing against the 'ashar advance payments to be made by tax-farmers on account of the taxes that were to be collected locally. Evidently, economic affairs were not running smoothly in the two sanjaqs of Jerusalem and Nablus: the bridge was built 18 years behind schedule.[13]

Transport between the different Transjordanian population centres was easier when it did not involve crossing the wadis or rivers in winter, and there were no intact bridges on the Yarmuk, Zarqa', Mujib or al-Hasa. On many occasions both people and animals were drowned while trying to cross these swollen waters during the winter season.

To acquire protection, travellers generally had to travel in groups to avoid being attacked by different tribes claiming a toll for the right of passage in their tribal territory,[14] or by ordinary highway robbers. Trade with the beduin tribes of the dira, as well as visiting tribes, had always taken place; in view of the large numbers of camels and donkeys owned by the tribes, they generally took care of transport as well. The settled population recognized that the beduins were better equipped to handle tribal disputes and to cross arid areas on the fringe of the desert.

These tribes were, and continued to be until the 1940s, important clients of the settled areas in Transjordan. They normally brought in their products of camels, sheep, wool, camel hair, samna and *jamid* (dried yoghurt) and exchanged them for their requirements of consumer articles, especially items of clothing and kitchen-ware. However, the most important requirement was grain, especially wheat. It was so important for them that they developed a phrase to describe the operation. When a clan moved westward in summer, or a beduin shaykh and a few of his kinsmen travelled west with a number of unloaded burden-camels, the word would go around that *hum rahu yiktalu* (they went to measure grain); their real purpose of course was to buy grain but the usage became so common that this expression, beduin in origin, became recognized by some governmental departments in the Arab world. During the second world war, centres where this operation took place in Iraq, such as al-Najaf, Karbala, 'Ayn al-Tamr, al-Samawa and al-Zubayr[15] were called Marakiz al-Iktiyal.

The production of cereals in Transjordan kept pace with the requirements of the slowly growing population until the 1870s, when the establishment of the Ottoman administration in al-Salt brought about greater stability and when the relative pacification of the tribes in al-Balqa' took place. It was then that farmers started moving eastwards and much larger areas of land were brought under cultivation. The crops produced were then much bigger and in excess of the requirements of the settled population and the nomadic tribes together. A surplus started to accumulate at the end of every harvest and farmers were forced to start seeking larger storage facilities. Generally these were old wells; they were cleaned and a layer of chaff was used as insulation between the seeds and the cold and sometimes humid sides of the wells. Normally wheat was kept for three years in the wells when prices were low or there was no demand for grain, but there were cases when it was stored for longer. When the wheat was brought out after such a lapse of time there was usually a loss of 5–10 per cent due to moisture or fungus, but this was generally compensated by the higher prices obtained for the produce, especially in winter.[16]

This old and inexpensive method of storage was still limited by economic factors, most important of which was the continuous need for cash or barter facilities. With the rise of yields, farmers encountered changing trends of consumption that were made easier by the improvements in transport and trade. They started looking around for new markets for their produce whose developing consumption needs could be met. Since Transjordanian towns were producers themselves and could satisfy the needs of their growing populations, the farmers in the new frontier areas had to look elsewhere.[17] Fortunately for them, larger numbers of beduin tribes were moving northwards to buy wheat during summer, and this specific trade may have accounted for 25 per cent of the whole Transjordanian crop.[18] But the real breakthrough came when the growth in the population of Palestine increased the demand for agricultural produce in general and for wheat in particular.[19] Wheat farmers in the Jabal 'Ajlun and Irbid areas started sending their surplus to Nazareth and Haifa and farmers in al-Balqa' started sending their surplus to Jerusalem and Nablus.[20] This trade was of great importance for the farmers since they not only managed to sell their produce for cash, but were also able to make some extra money from transport fees by sending the grain on their own animals during the off-season period just after the harvest.

As with other fields of activity during this period, there are no actual records available for reference. Both Palestine and Transjordan were under Ottoman rule and although the different parts belonged to different wilayat (governorates), there were no hindrances to travel and transport other than lack of security and highway robbery; and there are no records concerning customs or excise taxes as these were usually paid in the place where the goods were sold. The Abujaber papers of al-Yaduda do not include any records prior to the early years of the twentieth century. Although these are incomplete and damaged, it is still useful to refer to them; they are the nearest to our period of study and, as changes occurred only slowly, probably reflect earlier decades quite well.

The earliest documents are two receipts, both dated 6 August 1908, one for 400 bags and the other for 90 bags of wheat. It is stated that after receipt of those two quantities, the Abujabers were still in debt to the merchants Anton Ma'tuq and Partners in Jerusalem[21] for 25,692 piastres. The first receipt confirms that this account was then settled when the merchants bought from the Abujabers a two-year-old Arab mare for 100 French gold pounds (equivalent to nearly 10,000 piastres)[22] and received the balance in cash. The third earliest document is a letter addressed by Sa'd, who was then a student at St George's School in Jerusalem,[23] to his father Farhan, dated 26 August 1908. It is written on a letterhead of Nashashibi, Talil and Marar[24] and confirms that the camel caravan arrived safely and that the wheat was placed in the stores. It also states that all the harathin were paid in full and have therefore returned to their families.[25] This letter reveals that the commercial relationship that had existed with Anton Ma'tuq and Partners for many years was being brought, for some unknown reason, to an end, and that a new relationship was developing with the newly formed partnership of Mashashibi, Talil and Marar. Salim Marar, who must have been then around 29 years old, was a shaykh of the 'Uzayzat tribe of Madaba with whom the Abujabers had close ties. It is quite possible that the change of trading partners was due to his participation in the new company. This is confirmed by a letter from Farhan in Jerusalem to his son Sa'd, at al-Yaduda, dated 26 July 1911, which says that the wheat which arrived safely with the camel caravan was stored 'under the hands of Husayn Effendi al-Nashashibi, Mikha'il Talil and Salim Marar, as these are people who are well prepared for work. Any wheat shipped from now onwards is also to be directed to Husayn Effendi and partners.' It is interesting to note that with the returning unloaded caravan, supplies bought in Jerusalem were sent back to the families in al-Salt or to the households at al-Yaduda. In 1908 Sa'd sent back two large canisters full of *'araq* (an alcoholic drink distilled from grapes), 10 large melons, a crate of Samsun tobacco (produce of Samsun in Turkey) and 24 lemons. None of these items was available in Transjordan and their prices were not mentioned. In 1911 Farhan sent the following:

Article	**Quantity**	**Price**	**Total**
Sugar	4 bags	179	716
Rice-Egyptian	4 bags	211	844
Kerosene	20 crates	51	1,020
Olive oil	12 rutls +	18	217
Coffee	20 rutls +	28	568
Brooms	9		
Pepper	12 rutls		
Spices	12 rutls		
Cardamoms	3 'awqiya		34
Turmeric	3 'awqiya		
Black sesame	1 rutl		3
Cotton cloth	12 pikes		19
English rice	1 bag	160	160
			3,581 piastres

This amount would have been the equivalent of the price of three tons of wheat or the loads of 13 camels.

Having a trading account in Jerusalem did not, however, prevent the Abujabers from concluding transactions with other parties. During the 1911 season, Farhan sold 12,000 sa‘s of wheat to al-Khawaja Antun Farwaji, which was delivered as per the details appearing in table 6.1. The value of this transaction, or bazaar as it was called then, was nearly 600 French gold pounds. At the same time Farah sold to Antun Farwaji another quantity of wheat amounting to 10,026 sa‘s (60 tons) which was delivered between 10 and 30 July 1911. Simultaneously, another reasonably large quantity was being shipped on a consignment basis to another merchant in Jerusalem, Ahmad Tattan and son Husayn.[26] Like the previous shipment, this was carried mainly on the donkeys of three tribes of the Jerusalem area: al-Sawahrih, al-Ta‘amrih, and al-Diyasih (see table 6.2). The name of al-Qaysiya, a tribe of the Hebron area, appears once and some of the Abujaber camels were also employed to transfer 116 bags or 58 camel loads. The accounts for this consignment were settled by

Table 6.1 Quantities of wheat supplied by Farhan Abujaber to Antun Farwaji in Jerusalem during 1911 (average load small sack per donkey or 2 big sacks per camel)

Accompanying persons	Tribe	Sacks	Sa‘s	kg	Dates
12	Sawahra	67	978	5,868	30 July and 2 August
24	Ta’amra	87	1,242	7,452	31 July and 3 August
1	Diyasa	3	45	270	5 August
16	Sawahra	101	1,491	8,946	7 August
4	Diyasa	27	390	2,340	8 August
1	Sawahra	8	141	846	9 August
14	Sawahra	96	1,470	8,820	11 August
21	Ta’amra	97	1,359	8,154	14 August
8	Diyasa	35	528	3,168	18 August
14	Ta’amra	62	888	5,328	19 August
15	Ta’amra	72	1,038	6,228	20 August
9	Sawahra	42	684	4,104	21 August
13	Sawahra	75	1,098	6,588	22 August
–	Farhan’s camels — 18 loads	36	648	3,888	12 August
		808	12,000	72,000	within 24 days

Table 6.2 Quantities of wheat supplied by Farhan Abujaber to Ahmad Tattan in Jerusalem during 1911

Accompanying persons	Tribe	Sacks	Sa'‘s	kg	Date
7	Sawahra	49	726	4,356	11 August
2	Sawahra	16	258	1,548	22 August
18	Sawahra	122	1,869	11,214	25 August
7	Qaysya	17	333	1,998	14 September
16	Sawahra	114	1,737	10,422	11 & 14 September
9	Diyasa	47	666	3,996	4 September
5	Ta‘amra	24	336	2,016	6 September
11	Sawahra	81	1,227	7,362	13 September
5	Ta‘amra	25	366	2,196	15 September
8	Sawahra	40	597	3,582	22 October
	Farhan's camels — 10 camel loads	20	396	2,376	13 November
		555	8,511	51,066	within 52 days
Other products					
Sesame own camels		40	1,149		3 November
Hulba (Fenugreek) Diyasa		45	648		20 November
Hulba (Fenugreek) own camels		20	396		22 November

Tattan between 3 September and 12 November 1911, through payments made to third parties at the request of the consignors. The net amount, after all expenses, amounted to 52,182 piastres, equivalent to about 460 French gold pounds. The names of other grain merchants also appear in the records: ‘Umar Mustafa al-Daqqaq and Ishaq ‘Ali Hasna, ‘Abd Allah Badr, Muhammad ‘Ali al-‘Alami, and Hanna Francis Batatu.[27] There were also relationships with some of the institutions in Jerusalem. A consignment of 1,869 sa‘s of wheat (11.25 tons) was sent to the Frères de la doctrine chrétienne in Jerusalem on 22 August 1911. On 31 January 1913, Farhan Abujaber sent 100 tubba (2.5 tons) of wheat at the request of ‘Abd al-Rahman Effendi al-Budayri to the Municipal Council of Jerusalem.[28] On 27 October 1913 a receipt, signed by Nazarit Shahbarian of the Armenian patriarchate in Jerusalem, stated that 4,065 sa‘s of wheat (25 tons) had been well received. Gifts were normally not mentioned in these records as they were made in al-Yaduda, but on 23 July 1911 ‘Umar al-Daqqaq delivered to the Latin patriarchate two camel loads of wheat at the instructions of Farhan.[29]

Trade with the Nablus area from al-Yaduda must have been on a much smaller scale as only one letter was found from a merchant in Nablus; the signature could not be deciphered. The transaction involved a total amount of only 5,112 piastres. Trade with merchants in Amman started to develop hand in hand with the city's growth. Generally more was bought than sold there since the demand for cereals was very small; the Circassians in and around Amman were themselves farmers and had a surplus supply to sell. Two names appear often in the accounts. One is that of Hajj Khayru Effendi al-Sa‘udi, the head of an old family of Damascene origin, whose descendants are well-known government officials and merchants. The other is Yusuf Effendi ‘Asfur whose family came from Nablus via al-Salt. He served as Mayor of Amman and his descendants are a prominent family controlling one of the important business groups in the city.

This trade, although continuous and well-regulated by the standards of the day, had its difficulties and bad times. Shipments on consignment were generally stored at the expense of the consignor until the consignee, always a merchant in Jerusalem, found a willing buyer.[30] When this happened the Abujabers could only obtain minimum liquidity from the merchant since the few banks operational in Jerusalem did not allow credit to people who were not resident in Jerusalem.[31] Another difficulty was pilferage or shortage of contents. ‘Umar al-Daqqaq and ‘Ali Hasna, in a letter to Farah at al-Yaduda dated 16 Jumada al-Akhira 1333 H/1914 AD, complained that the expected 72 tubba of barley was found to weigh only 66. In another letter six months later they complained that the expected 538 sa‘s of wheat were found to weigh only 496 and that they had heard that the caravan attendants had sold small quantities in the market without telling them. In most cases such shortages were claimed and collected from the transport contractors or the caravan attendants in spite of the unpleasantness involved.

Transport costs were relatively high. Bridge tolls were at the rate of 8–10 piastres per camel and 2–3 piastres per donkey. To this was added the expenses of the caravan attendants and a certain *ardiya* (demurrage) which was probably charged at the bridge. This amounted to nearly 3 piastres per camel load and 1 piastre per donkey load. The main item was the fare paid to the animal owners; this naturally varied according to supply and demand. A statement from Tattan to Farhan at al-Yaduda on 30 Dhu al-Qa'da 1329 H/1911 AD mentions the fare for 23 camel loads as 617 piastres and the fare for 40 donkey loads as 758 piastres.[32] Thus the total cost was nearly 40 piastres per camel load and nearly 23 piastres per donkey load, which was between 12 per cent and 15 per cent of the value of the cereals carried. Although it was relatively cheaper to transport by camels they were not always available and donkeys also had to be used. The difference is highlighted by Tattan in a letter to Farah dated 19 Ramadan 1331 H/1912 AD when he advises the Abujabers to start loading with camel owners for transport, as they were more reasonable than the donkey owners. This piece of advice was heeded and from that year the records show larger quantities being carried on camels. The Bani Sakhr and al-Hawarna began to appear in the lists as carriers: the former transported 170 bags and the latter 78 bags during 1331 H/1911 AD.

Cost accountancy was not one of the worries for the producers of cereals in Transjordan. They had a crop and to turn it into cash or purchasing power through barter was their final aim. They were therefore willing to sell it on the spot or to transport it to the markets where it fetched an acceptable price. Fortunately for them, they were not handling a perishable product: they could store their crops for two or three years. This, it is true, would mean hardship and the inability to buy the merchandise they needed; but they did have an advantage, although it may not have appeared to them as such at the time. The population was growing at an exceedingly high rate both in Palestine and to a lesser degree in Transjordan, and they happened to be producing what was most needed by many of them. A new class of consumers was gathering in new urban centres that was neither able nor willing to produce its own requirements of food, and the Transjordan producers moved in to fill the gap.[33]

7 Taxation of Agricultural Produce

Organization

The fiscal administration of wilaya Suriya province of Syria was a relatively simple system that was directly linked to the *wali* (governor) in the province's capital, Damascus. The official yearbook (*Salname*) for the province of Syria in 1879 shows that the wali at that date was Ahmad Hamdi Pasha, who appears to have been deeply interested in the affairs of the area placed under his care.[1] The fiscal affairs of the whole governorate were at that time in the hands of the *daftardar* (chief financial officer of a province) Riza Effendi who also resided in Damascus. He was in charge of collecting the revenue and defraying the expenditure. The wilaya was divided into the eight *liwa's* or *sanjaqs* (sub-provinces) of Damascus, Acre, Balqa'[2] (which included Jabal Nablus and the district of al-Balqa' in Transjordan), Hawran (which included Jabal 'Ajlun in Transjordan), Hama, Bayrut, Tarablus and Ladhiqiya. In each of these sanjaqs there was a *muhasibji* or *mudir muhasaba* (fiscal director) who resided in the capital of the sanjaq. Immediately under the eight directors there was a *mudir mal* (fiscal officer) in each of the thirty important towns in the wilaya of Damascus, each of which was the capital of a *qada'* (district). The records do not reveal that there were any fiscal administrators stationed in the smallest Ottoman administrative units, *nahiyat* (sing. *nahiya*, sub-districts).[3]

In 1879 Transjordan had only two fiscal officers, Iskandar

Kassab Effendi in al-Salt and Salim Farkuh Effendi in ‘Ajlun.[4] Both these posts had been in operation for some time and we have it on record that in 1288 H/1871 AD the mudir mal in ‘Ajlun was Ibrahim Mawla Effendi while the mudir mal in al-Salt was Asad Frayha Effendi.[5] Although al-Karak did not have any real Ottoman administration or presence until 1894, it was still listed as a qada’. The *qa’immaqam* (governor of a qada’) is mentioned as Muhammad Majali Agha who was, anyway, the leader of al-Karak, and he had only one *katib* or scribe.[6] The other qada’, Ma’an, had an impressive military presence in the form of a few army units, but two vacant posts of qa’immaqam and mudir mal. This may be explained by the fact that Ma‘an’s importance lay much more in its being the major station on the yearly pilgrimage route than a centre for the collection of taxes that were anyway of not much importance.[7]

This pattern was the same as that applied in the different areas of wilaya Suriya; the list which appears on the following page shows the different liwa’s and qada’s in the year 1879 and gives the names of the government officials who were involved with the Transjordan countryside in administrative posts concerned with executive and fiscal functions.

Taxes and their types

Transjordan had one major agricultural tax, the ‘ushr (tithe), generally known as *mal miri*. *Miri* means that which pertains to the amir and *mal* is money or tax; thus it was the tax owing to the authority or state. As its name in Arabic implies, the ‘ushr was originally equal to one-tenth of the produce of the soil. The amount it represented, however, had been increased from time to time by Ottoman governments for revenue purposes. At the time of the British occupation of Palestine and Transjordan in 1917–18, ‘it was collected at the rate of 12.5 per cent. Crops were assessed on the threshing floor or in the field and the tithe was collected from the cultivators.’[8]

The increase in the rate of the ‘ushr tax from 10 per cent to 12.5 per cent took place gradually during the period under study. It came about as a direct result of the financial difficulties faced by the Ottoman Empire. Syrian sources

Administrative and fiscal organization of the Transjordanian parts of the Ottoman province of Syria in 1879

(Note: names are of those who held office in that year.)

1 Province (*wilaya*) of Syria.
  Governor (*wali*): Ahmad Hamdi Pasha.
  Chief financial officer (*daftardar*): Riza Effendi.

2 Sub-provinces (*sanjaqs*). Of the 8 *sanjaqs* in the province, two included parts of Transjordan: Balqa' and Hawran.
  Balqa': sub-governor (*mutasarrif*): Khalil Bey.
    chief financial officer (*muhasibji*): Zayn al-'Abidin.
  Hawran: (*mutasarrif*): Mahmud Pasha.
    Muhasibji: Hilmi Effendi.

3 Districts (*qada's*). The Transjordanian parts of the *sanjaq* of Balqa' were included in the *qada'* of al-Salt; those of Hawran in the *qada'* of 'Ajlun.
  al-Salt: administrative officer (*qa'immaqam*): Muhammad 'Ali Effendi.
    financial officer (*mudir mal*): Iskandar Kassab Effendi.
  'Ajlun: *qa'immaqam*: Husayn Bey.
    mudir mal: Salim Farkuh Effendi.

4 Sub-districts (*nahiyas*): the *qada'* of al-Salt had two *nahiyas*, Jiza and Ma'an; that of 'Ajlun was not sub-divided into *nahiyas*.
  Jiza: administrative officer (*mudir*): Sattam al-Fayiz.
  Ma'an: *mudir*: Shukri Effendi.
  There were no financial officers at this level.

mentioned that 0.25 per cent was added in 1878 for the payment of the war indemnity to Russia. Another 1 per cent was added for the creation of an agricultural bank in 1885 and an extra 0.5 per cent for education in the Empire. The last addition of 0.5 per cent was made for the supply of military equipment to the armed forces.[9] There is confusion in the sources and the figures given here do not account for 2.5 per cent. Another source writing about taxation in the 1900s in Iraq shows a similar confusion there: the tax rate was raised in 1883 by 1 per cent allocated to the Agricultural Bank and 0.5 per cent to public education; in 1897 a further addition of 0.5

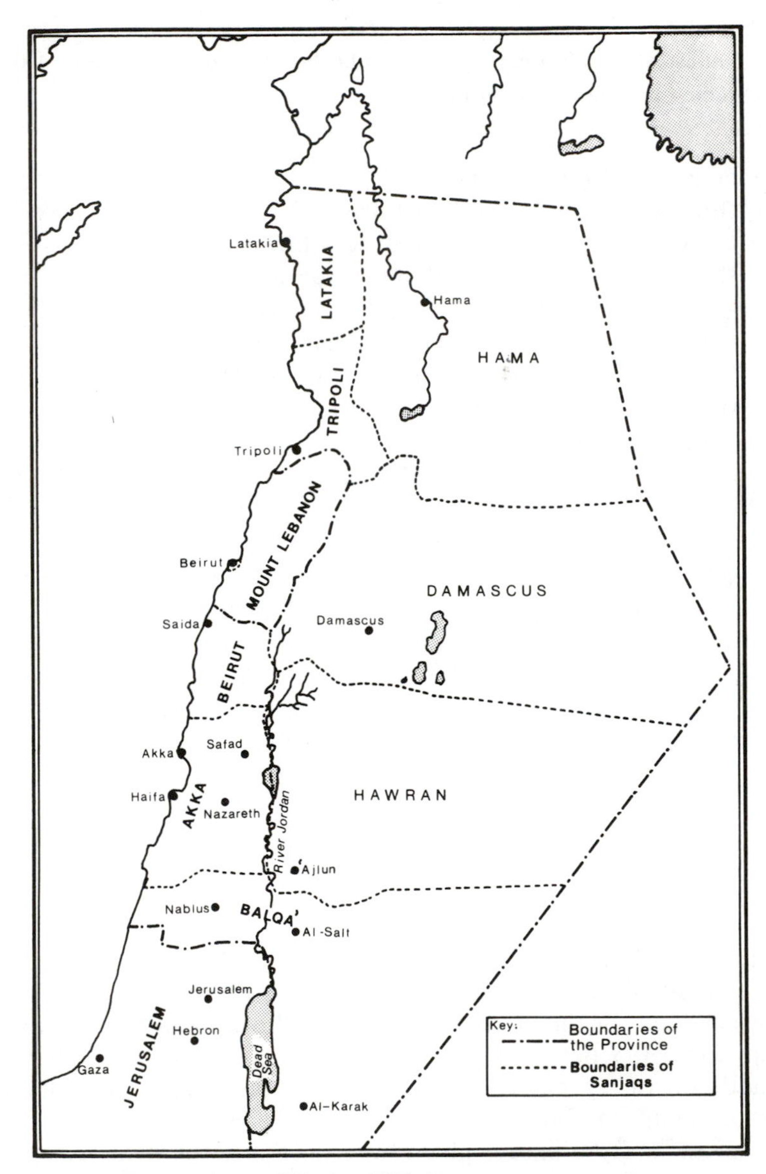

Map 7.1 The province of Syria, 1880–1
Note: Mount Lebanon and the Sanjaq of Jerusalem had a special status.
Sources: A. Schölch, *Palästina in Umbruch* (Stuttgart, 1986); *Salname-i Vilayet-i Suriye* (Damascus 1298 AH/1880–1 AD).

per cent was made; and in 1900 one of 0.63 per cent for military purposes. As a result the total had risen to 12.63 per cent.[10] It is certain, however, that during the latter years of Ottoman rule, prior to 1918, the farmers of Transjordan were paying the ‘ushr one-eighth of the crop or 12.5 per cent.

Another tax, the *werko*, sometimes also written *vergo*, was also of importance. It was in effect a house and land tax.[11] It was levied on immovable property based on capital value and varied from 4 per thousand on miri land to 10 per thousand on *mulk* property. (Under the Land Law of 1858, land in the Ottoman Empire was divided into five classes: they included mulk land that was possessed in full ownership and therefore held privately; and miri or state land which comprised arable fields, meadows, pasture land, woodland and the like, the enjoyment of which was granted by the government, while the legal ownership was vested in the treasury.)[12]

Due to the lack of documents it has not been possible to ascertain how the werko was collected and at what rates. No receipts were found in al-Yaduda papers, even for payments on the Abujaber property in al-Salt, although elders of the town vaguely remember paying such taxes. This may be because they were lost when most of the houses in al-Salt were sacked towards the end of the first world war. This happened when the town was evacuated by the British forces for military reasons. The population, having supported the British military effort within the framework of the Arab revolt against the Turks for independence, became apprehensive of retribution, panicked and within hours evacuated the town in their turn. Over 5,000 people continued their flight to Jerusalem, 80 kilometres to the west, while the rest spread out in the countryside.[13] That same night the town was sacked by the returning Turkish troops and marauding parties from the villages and encampments in the countryside.

A third tax of importance was the one levied on animals, usually called *aghnam*. Sources give different figures for different districts in the Ottoman Empire but it seems that in Transjordan the basic figure of 4 piastres per sheep or goat and 10 piastres per camel or buffalo specified during 1255 H/ 1839 AD was still being applied in 1917.[14]

The Transjordanians were generally not happy to pay

taxes, and in many instances resorted to armed resistance. This dissatisfaction, together with the inefficiency and corruption in administrative and military circles, may have been the reason for the non-payment of taxes levied in other parts of the Ottoman Empire, such as *ferde* (personal tax), *salyane* (annual tribute) or *'askariya* (military tax). Elders do not recall that these were paid by their fathers and al-Yaduda papers do not provide any documents to that effect.

This situation, however, is the subject of conflicting reports in books of the time and must have been different from place to place, depending on the strength of the Ottoman administration in any given area. Further, it must have been directly influenced by the gradual consolidation of governmental presence in the different districts after a governor was instated in 'Ajlun in 1851. Oliphant, who wrote in 1879, confirmed that the population did not pay taxes on their crops.[15] Another traveller, Schumacher, after his visit to the Jawlan, not far from Northern Transjordan, wrote that the people were taxed instead of military service at the rate of 120 piastres per tent of five persons. (1 gold Napoleon = 90 piastres.)[16] A more recent reference is that of Musil, who visited during the first years of the twentieth century. He confirmed that the tribes of Madaba, who were all Christian, were paying taxes, at the following rates per family:[17]

4	majidis	(80 piastres)	per faddan (pair of oxen)
1/5	majidi	(4 piastres)	per sheep or goat
1	majidi	(20 piastres)	'Askariya (military tax)

It is important however to mention that, although the writer mentioned 14 tribes in the area of al-Balqa' and al-Karak, the taxes were only mentioned once when he was speaking about the tribes of Madaba. The fact that these people were new settlers after their departure from al-Karak may explain why they were alone on this disciplined course. Most probably they were paying taxes in order to claim government support and protection should their rights to the land be contested by the formidable Bani Sakhr.

An agricultural tax that was of some importance to the people of al-Salt was that levied on produce when transported to the markets of Jerusalem or Nablus. This was rather high and was paid in Jerusalem at the rate of one Turkish lira per load of raisins, jamid, summaq or *rumman* (pomegranates). The same taxes were applied to the produce of the ‘Ajlun area.[18] Although it is not specified, it is most probable that the load meant that of a camel and that lower taxes were paid on the loads of horses, mules and donkeys.

Taxes in the form of temporary collections seem to have been made every now and then. A receipt found among al-Yaduda’s papers was one given to Farhan Abujaber and brothers by the mudir of Nahiya ‘Amman on 17 June 1317 H/1899 AD. It is interesting because it involved the payment of 12 majidis (nearly 2.5 gold pounds) for *i‘ana akhshab* (wood aid) for the Hijaz railway which was demanded from Qada’ al-Salt. This must have been a temporary tax that probably lasted until the construction of the railway was completed.[19]

Tax assessment procedures and collection operations

Prior to the year 1329 H/1911 AD, al-Yaduda papers do not provide any records of how the taxes were assessed; nor do they include any receipts of tax payments. Nevertheless, there is evidence that Salih Abujaber was already paying taxes in 1879.[20] The papers do include 27 receipts for amounts which were paid between 1316 H/1889 AD and 1328 H/1910 AD to the original owners of the lands that were bought over the years by Salih and his three sons. These payments were made *badal a‘shar* (in lieu of the tithes) and there is no conclusive evidence that any money was paid during these years directly to the government. Although one receipt signed on 29 Sha‘ban 1322 H/1904 AD by the *mukhtar* (headman) of al-Zififa (a tribe bordering al-Yaduda on the north) states that the 6 majidi riyals received (equivalent to nearly 1.5 gold pounds) were paid to the *khazina ‘amira* (treasury), it is safe to assume that because the original owners of these lands were members of beduin tribes, they were unofficially exempted from such payments. It is also of interest to note that some of these lands were already registered in the name of the Abujabers, and yet the value of the tax was paid to the original owners, the shaykhs of al-‘Ajarma tribe. The same applies to the lands bought from the sons of Sattam ibn Fayiz, shaykhs of the Bani Sakhr tribe, and known as al-Nu'ajiya.[21] The receipt is signed by Shaykh Fayiz, the second of Sattam's sons, and involves 20 majidi riyals, or about 5 Ottoman gold pounds. As Sattam had eight sons, the Abujabers paid during 1324 H/1907 AD badal a'shar the sum of nearly 40 gold pounds on al-Nu'ajiya. This was a reasonable tax to pay for nearly 4,000 dunums (400 hectares) of the best agricultural land in the area. Payment of amounts of money in lieu of taxes by the cultivators of the land to the original beduin owners was made permissible by a decision at the governors level in Damascus. Probably it was a temporary concession that was extended in the hope of allowing a larger number of cultivators to spread out in the areas controlled by the beduins.

For the year 1329 H/1911 AD al-Yaduda papers include 59 official receipts showing the delivery in kind of the taxes due from the agricultural crop produced on the estate during the month of July 1329 H/1911 AD to the government authorities in al-Salt. This development coincides with the determination of the Ottoman authorities to collect arms, register property and lands, subject people to more systematic taxation, and to organize a census that was considered an introduction of the conscription system. This policy led to the revolt in Jabal Druze in 1910 and in al-Karak in 1911.[22] The administration in the frontier of settlement had started at long last to collect taxes directly from those who owed them without giving any consideration to the beduin influence.

One of the 59 receipts for 1911, dated 16 July, refers to a previous ruling by the *Majlis Idara* (administrative council) in al-Salt. The assessment, having been agreed by the owners of al-Yaduda, was also held as final by the council. The total quantity of cereals for that year's tax amounted to:

	kg
Wheat	26,787
Barley	21,018
Lentils	4,413
Vetches	684
Beans	1,005
Hulba	411
Total	54,318

The following year, 1912, started well with a pleasant letter from ‘Isa ‘Abd allah al Hamarna[23] informing the three Abujaber brothers that he and his partner, Hanna Effendi Farah,[24] the tax-farmers of al-Yaduda, had delegated Salim Zu‘mut[25] to take delivery of the a‘shar against official receipts signed and stamped by him. He was to be assisted in his work by three colleagues from his clan in al-Salt. During August, cereals for the taxes were delivered against 7 receipts and they amounted to:

	kg
Wheat	36,684
Barley	38,340
Lentils	9,882
Vetches	108
Beans	684
Hulba	888
Total	86,586

On 20 November 1912, Salim Zu‘mut and his colleagues submitted a report to the tax-farmers confirming that they had been entrusted to collect the taxes from al-Yaduda, had proceeded with their work and collected what was due in taxes. However, they had been unable to collect taxes on the fields that the Abujabers, through collusion with another tax-farmer, claimed belonged to a neighbouring village. They asserted that the extra taxes still to be settled were:

	kg
Wheat	45,270
Barley	23,029
Vetches	5,634
Sesame	2,928
Farik[26]	645
Total	77,506

Although there were no extras required for lentils, beans and hulba, the increase was nearly 90 per cent of the quantities already delivered to the tax collectors in August.

On the strength of this report the tax-farmers brought the matter to the knowledge of the qa’immaqam of al-Salt on 13

December 1912. The amount claimed for sesame was increased to 7,122 instead of 2,928 kilograms. *Daftars* of taxes were submitted and witnesses were called in. Although the letter sent by the qa'immaqam to the Abujabers is not among the papers, it seems to have been dated 4 December 1912 and was a request for payment as per the report. Within a few days the owners of al-Yaduda replied to the qa'immaqam, charging the tax collectors with fraud and falsification of documents. Their point was that taxes were assessed by the authorities and agreed to by the tax-farmers who took delivery of the quantities of cereals in August 1912. They were not willing to pay any extra taxes to the tax-farmers and they were certain that justice would not permit any new assessment to take place three months after the crop had been removed from the threshing floor. The Fiscal Bureau in Damascus, to which the case was probably referred, must have felt, as they did in a previous case, that since the tax-farmers were encouraged and assured that influential taxpayers would be made to raise their payments by 10 per cent, the taxpayers were to co-operate, provided that they be given the opportunity to explain their point of view in case of objection.[27] Unfortunately there are no documents which reveal what happened, but a document signed and stamped on 20 January 1913 by Hanna Ibn Farah admits that a settlement was reached and that as of that date none of the parties had any claim against any other. Furthermore, the court action brought by 'Isa al-Hamarna and Hanna Ibn Farah was to be considered cancelled and without any effect.

The following year provides us with another case in another area of agricultural activity. The document, found among al-Yaduda papers, is dated 15 January 1913. It is a complaint brought by a group of farmers in the valley of al-Salt and submitted to the inspector of property in the region. They start their petition by saying that 'the presence of governments is only there to ensure the safeguard of the rights of the people and to remove from among them any despotic authority.' They claim that the farmers at Batna[28] had managed to transfer most of the tax load from their fields to those of the applicants and that furthermore their crops had been eaten by locusts. When they protested, they were surprised to find themselves hosts to a detachment of cavalry who were behaving most unpleasantly and were clearly taking the side of the farmers at Batna. They charge fraud in the *daftars* (registers) and complain that the cavalry, after being withdrawn for a short while, came again and resorted to their old acts, including putting some of the farmers in gaol. Evidently the complaints were so strong and loud that the case was brought to the attention of the qa'immaqam who ordered that it be considered by the mudir mal of al-Salt; on the same day the mudir mal decided to turn the whole case over to the Majlis Idara for reconsideration. Most likely, like all other cases reconsidered by the council, the dispute between the farmers was solved through a compromise.

Another picture can be found in the records of al-Yaduda that same year. During May, the *multazims* (tax-farmers) Saydu al-Kurdi and his partners[29] addressed a very kind letter to Farhan Effendi Abujaber and his brothers, starting with the ingratiating words, *Janab hadarat al-ikhwan al-mukarramun* (the exalted dignitaries and highly respected brothers).[30] They stated that the tax tender for al-Yaduda had been awarded to them for the year and for that purpose they had delegated one of them, a certain Isma'il Bey,[31] to receive the taxes against receipts signed by him. Negotiations must have then started and after many sittings, debates and arguments, an agreement was arrived at whereby Farhan, on behalf of al-Yaduda's owners, undertook to deliver cereals for the amount of that year's miri amounting to 100,000 piastres (1,000 Ottoman gold pounds). The cereals were to be approximately 75 per cent wheat, 20 per cent barley and 5 per cent vetches; quantities were to be according to specifications set down by the authorities and delivered within two months of the agreement date to the government's cereal stores either at Zizia or at Amman.[32] Should there be a delay, a fine of 200 gold pounds would become due. Furthermore, Farhan undertook to settle all expenses of controllers, stamps, fees for *iltizam* (tax contract) certification and register fees. This document was dated 20 July 1332 H/1913 AD and on 1 November Saydu signed on the back of it a receipt stating that the cereals had been delivered to the government stores at Zizia[33] on time and that Farhan had fully settled his obligation for that amount. Saydu received 45 Ottoman gold pounds to cover expenses for the tax collection at al-Yaduda for the year 1332 H. Farhan further paid, against a separate receipt, the sum of 450 piastres, being the salary of the controllers who stayed at al-Yaduda during that period as the uninvited guests of the estate.

When the information provided by these documents is considered, it becomes clear that 1332 H must have been a good year. A total of nearly 100 tons of cereals, or 1,000 gold pounds in value, was settled in taxes, and as this was only 12.5 per cent of the estimated crop, it means that al-Yaduda's owners admitted to a total crop that year of nearly 8,000 gold pounds. It was probably 30–40 per cent more. Also interesting is that it gives an indication of the amounts of the different crops being cultivated, although the proportions were not always the same. The ‘ushr was generally based on an estimate of the harvest, and for the following year, 1333 H/1914 AD, was calculated at 45,175 kilograms of wheat and 23,000 kilograms of barley, plus expenses, during a public auction held for the sales of this iltizam. The estimated crop in general was lower than in 1913 and the proportion between wheat and barley was 66 to 34 instead of 80 to 20. This may have been the result of planning on the part of the taxpayers: by declaring a larger crop of barley and a smaller crop of wheat than the actual, they stood to save a good amount of money. The saving could be around 50 per cent of the price difference since wheat prices were much higher than barley prices.

Generally, people in the settled areas of the north and the middle of Transjordan started paying some taxes after the administration was established in ‘Ajlun in 1851 and in al-Salt in 1867. Nevertheless, taxes were a new and unpleasant development and were very unpopular with the people. This attitude prevailed for many years; a true picture of how people felt is contained in a letter sent by Frayh Abujaber to his older brother, Farhan, on 1 Jumada al-Thani 1317 H/7 October 1899, which reads as follows: ‘As to *al- a‘shar* [taxes on crops] Quwaytin Agha [son of ‘Aqila Agha] has arrived from al-Karak for the purpose of collecting al-a‘shar from the people of al-Salt. Up till now nobody has paid. May the

Almighty *remove this epidemic* in good time and may it all end well.'[34]

The Ottomans appear not to have had any fiscal plan for the area under study. Their tax-collection attempts were generally haphazard and could not, therefore, have brought in the desired results. As early as 1288 H/1871 AD petitions were made against the governors and the multazims alike. One such complaint was submitted to the Porte in Istanbul by people in the area of al-Karak; the Porte was induced to issue orders to the wali of Suriya to enquire into the matter.[35] It must, however, be stated that this haphazard approach was probably forced on the administration because, in some parts of Transjordan, agriculture until the 1890s was neither as established nor as prosperous as it was in the Hawran and its neighbouring district al-Jaydur. Tax collection in these more settled areas, even as early as 1261 H/1845 AD, was being carried out on the basis of mal miri or a specific rate per faddan. During the Egyptian period (1831–41) this was 346 piastres in Hawran and 321 piastres in al-Jaydur; it was raised by the Ottoman administration after the Egyptian withdrawal in 1841 to 406 and 365 piastres respectively.[36] Had such a system been applied in Transjordan, there is no doubt that it would have removed many of the complaints against over-assessment and corruption and given agriculture an opportunity to develop at a faster pace than it actually did.

It is not possible to give a full account of how the system worked at the start of the reinstatement of the Ottoman administration as the official documents for this period could not be located.[37] However, a reasonably clear picture can be obtained from papers of al-Yaduda relating to the first two decades of the twentieth century. These are as yet the oldest documents available to us for that period, and in view of the slow pace at which the Ottoman administration developed, it is safe to assume that the procedures in the documents were already being applied during the last two decades of the nineteenth century. It is important to mention that during these four decades Transjordan continued to be the southern part of the governorate of Damascus, in which the administration's main purpose was to consolidate the authority of the Ottoman Empire in these outlying parts.[38] The governor was more concerned to bring the countryside in Transjordan under control than to develop in it new administrative and fiscal systems.

It should also be borne in mind that al-Yaduda's papers are neither complete for any one year nor continuous for any number of years. Until around 1900, conditions were unstable and insecure and the second decade of the twentieth century witnessed all the anxiety of the first world war.

Amounts of taxes in Transjordan

From the documents available, it has not been possible to obtain figures of the total value of taxes in the area of study. Probably one day a daftar or report will be found in Nablus or Damascus which will give lists of taxes levied and collected for a certain year in every administrative unit in wilaya al-Sham. There is, however, some information about the revenue of wilaya Suriya in an article published in 1879; the authors give useful data about the district of Damascus in which agriculture was the chief source of revenue, as was the case in the whole province.[39] The official estimate for the revenue of the wilaya in 1873 was the equivalent of £S2,381,255, while the actual amount received by the administration was only £S629,337, or a mere 26.4 per cent of the total. This unhealthy state of affairs is attributed to depopulation as a result of the pressures by usurers and tax gatherers, and to a lesser degree to the 'wild Arabs'. Writing about the deserted villages, or khirab, was common at this time and by 'wild Arabs' the authors must have meant the nomadic tribes of the Syrian desert who were in the habit of exacting the khawa (tribute) from villagers living in the settled areas on the fringe of the desert. To demonstrate the importance of the yearly pilgrimage to the Holy Shrine, the authors mentioned that the yearly cost to the wilaya of Damascus for the running of the Hajj caravan amounted to the equivalent of £S42,575 in 1876 and to £S39,091 in 1877, which is nearly 6.5 per cent of the total yearly amount received by the administration.

Although no mention is made of Transjordan in the article, it is safe to assume that conditions there around 1880 did not differ much from those in the countryside around Damascus.

An interesting report is given by the author-traveller Laurence Oliphant. In 1880 he wrote that the area of ‘Ajlun, the northern district of Transjordan, some 40 miles by 25 miles, with 15 deserted villages out of 75 and a population of under 20,000 people, rendered a revenue that was not more than £S7,000.[40] A considerable proportion of this amount was paid by nomadic Arabs as *widy* (tax on animals).[41] The Bani Sakhr refused to pay taxes on their animals[42] and, having free access to Badiya al-Sham, probably went on their way without paying anything. Al-Salt is also mentioned by Oliphant; he confirms that its revenue was only £S1,000, ‘which was far short of the proper proportion’.[43]

In al-Karak, prior to 1880, attempts were made every now and then by the Ottoman authorities to collect taxes, but these were unsuccessful. Such an attempt was made by Suraya Pasha, the governor of Jerusalem, when in 1861 he dispatched an expedition to al-Karak which succeeded only in collecting 140 purses (a little more than £S500).[44] Another famous expedition was led by Hulu Pasha al-‘Abid, the governor of Nablus, in 1863,[45] against the ‘Adwan.[46] Between 1861 and 1876, at least four more expeditions or attempts to collect taxes were made in al-Karak and again they do not seem to have been successful. These were led by ‘Aqila Agha who insisted on receiving four lambs from every household, while Huṣayn Buzu accepted 25 piastres per family. The third ended in failure and the fourth, led by Salih Abujaber, collected wheat for taxes; this was stored at the house of the shaykh and could not be later removed as he had appropriated it.[47]

The report of Ahmad Hamdi Pasha, referred to at the start of this chapter, provides some information, but only on the sanjaq level. The revenue for sanjaq al-Balqa’, which included the areas of Nablus and the middle of Transjordan, amounted to a total of 7,637,362 piastres, of which the two main items were the werko, 2,072,885, and a‘shar, 4,409,273. In comparison, Hawran, including ‘Ajlun, showed a revenue totalling 2,752,663 from werko and 1,509,695 from a‘shar. Expenditures other than for salaries are not shown and the total figures for the final accounts are not added up. Only the allocation for the Hajj caravan that year appears; it was 3,700,000 piastres and was listed as an item of expenditure for the whole governorate.

Another interesting report is that presented by Musil a few years after the Ottomans dispatched an expeditionary force to al-Karak, and the population in the southern part of Transjordan started paying taxes on a more or less regular basis. He attempted to put on record the information he could gather in this regard and states that three shaykhs of the Majali clans were exempt from payment of taxes.[48] Probably this was allowed by the authorities in recognition of their services and status as leaders. The amounts that had to be paid by the different clans and villages were given as follows, in piastres:

Clans in al-Karak		
	Christians	64,000
	Gharaba	80,000
	Halasa	20,000
	Majali	12,000
	Sharaqa	83,000
Clans in the district		
	Hamayda/al-Karak	60,000
	Hajaya	13,500
	Hamayda/al Hasa	80,000
	Nu'aymat	30,000
	al-'Amr	32,000
	Hamayda/al-Jabal	80,000
Villages in the district		
	Kathraba	60,000
	al-'Iraq	25,000
	Khanzira	40,000
	Total of taxes	679,500 piastres

which is approximately 6,800 Ottoman gold pounds.[49]

In conclusion, taxes in Transjordan were a great cause of concern for the authorities and the population alike during the second half of the nineteenth century. The haphazard nature of their assessment and collection provided wide scope for mismanagement and corruption. The authorities tried to exact the most and the taxpayers tried to pay the least, and officials and their friends among the tax-farmers were active all the time in seizing opportunities to make money for themselves. No wonder therefore that the people considered tax collection and all that went with it – injustice, mistreatment and imprisonment – an epidemic they sincerely prayed would cease. Unfortunately for them, however, these difficult conditions not only persisted at the start of the twentieth century but were made even worse by the advent of the first world war. It was not until the 1920s[50] that the hated words, assessment and multazimun al-a'shar, disappeared from the vocabulary of the day and were replaced by fixed taxes which provided no opportunity for mistreatment or corruption.[51]

PART THREE

Case Studies of Agricultural Ventures

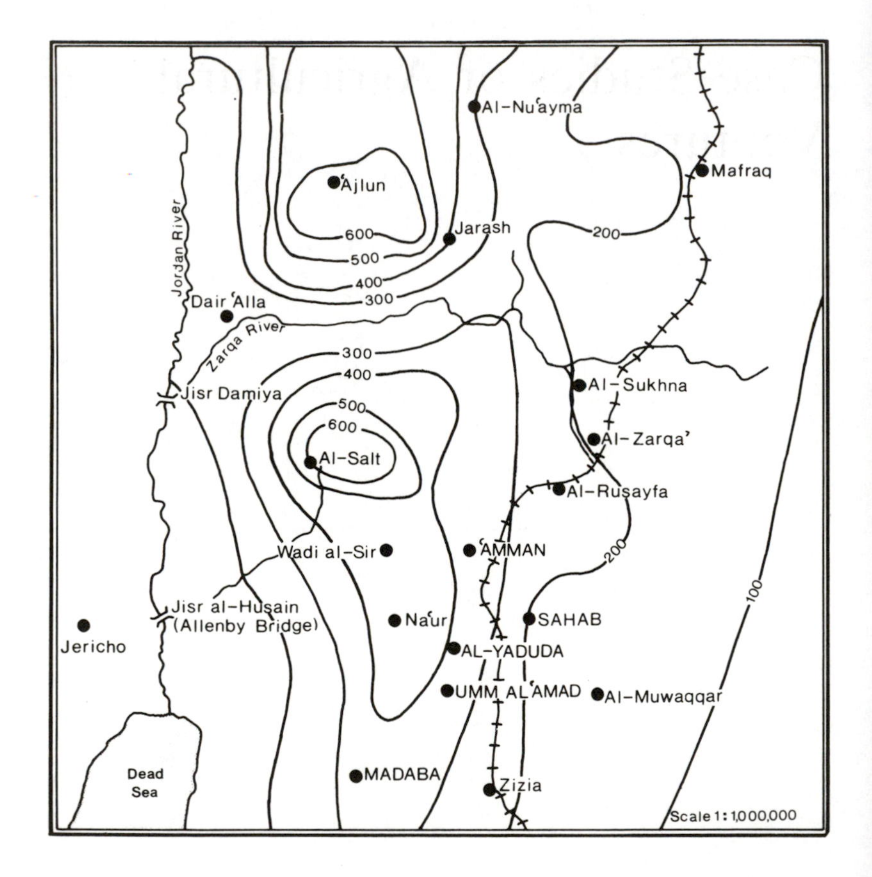

Map of case studies: situation and rainfall averages

Source: Mean Rainfall Map 1937–1961.

Introduction

The study of the development of agriculture in south-eastern Bilad al-Sham during the nineteenth century will not be complete unless it also involves the study of living conditions of farmers and settlers who migrated into the new frontier area. Generally they came from the settled mountainous area to the west, but there were also among them those who were originally beduins; those who came from foreign lands; and those who moved northwards from the south. Many of them, and especially those who came from outside the region, such as the Circassians and Chichen, had to fight hard against the adverse conditions that frequently confronted them in their new settlements. Yet there were very few, it seems now, who wished to change their new style of life. The ownership of land in the new frontier area was in itself a sufficient reward.

The six case studies presented here involve new settlements in the north and middle of Transjordan. Of the other parts of Transjordan, Al-Ghawr, or Jordan Valley, was occupied by different small tribal groups and their slave followers at the beginning of the nineteenth century (when this study starts) and was already under cultivation on a small scale. Objectively speaking, therefore, the Jordan Valley did not contribute to the development of agriculture in Transjordan during the nineteenth century and it was only in the 1930s that people started to appreciate the importance of the economic potential in the naturally rich agricultural zone. It is sufficient to say that, prior to that period, the Jordan Valley provided for the Jordanians good pasture during the winter months every year and only supported, through its agricultural produce, the small population that either lived in it or lived off it. The former were the Ghawarna groups who lived in it continuously; the latter were those tribes who owned land but were only ready to stay in it during the winter and spring months, thus avoiding malaria and other diseases. The number of the whole population in 1900 could not have been more than a few thousand by any count and their produce is not mentioned in any records as being of any great significance.

The development of agriculture in al-Karak and southern areas, on the other hand, was greatly hampered by the lack of stability and security that are essential for human activity to flourish. The spread into the eastern districts of the south, therefore, did not happen until after the start of the twentieth century and as such is beyond the scope of this study. Pastoral activity in the south tended, however, to compensate for the economic deficiency resulting from the absence of large-scale agriculture and the area, especially al-Karak, has been renowned for its large flocks of sheep and goats. This undoubtedly brought about the economic balance under which the relatively small population was able to survive until better days.

The different cases of settlement that are studied at length in this research are those that had a real impact on the development of agriculture in Transjordan during the nineteenth century. An attempt has been made to show how each of these settlements tried to meet the new requirements made of it by the agricultural pursuit it chose to follow. Also, a description of life among the newcomers, the beduins-turned-farmers and the settlers from foreign lands, has been included. They were different from each other in many respects and each group had different problems to solve. Nevertheless, disparate as they were, they had one thing in common. Every one of these six communities, as well as the other smaller ventures that are not specifically studied here, had the will and determination to start a new way of life in dry farming areas on the fringe of the desert. Basically they cultivated cereals, mainly wheat and barley, and to them pastoralism was second in importance and only a means of supporting the income they already derived from farming crops. They were the new farmers who accepted a challenge and were there to stay on the virgin land that they cultivated after it had been neglected for centuries. The geographical location of each of these six sites appears on the map of case studies (p. 130). With the exception of Sahab all of them fall within the annual rainfall zone of 300–400 mm, which guaranteed for their settlers the annual crops they so badly needed.

8 Al-Yaduda or Khirba Abujaber: the Pioneering Spirit

The oldest of the settlements in the middle part of Transjordan was that founded by two pioneering brothers, Ibrahim and Salih, sons of Nasir Abujaber who was probably born in Nazareth around 1775. This family, Latin Christian in faith, was originally from the southern part of Transjordan, according to family sources; their migration via Hawran to Nazareth could have taken place some two hundred years earlier.[1]

In 1815 the Abujaber family sold a house[2] to Michael Qa'war, the grandfather of the first Protestant Arab priest to be ordained in Palestine,[3] and probably in that year Nasir migrated to Nablus seeking a better career. After that we hear of his marriage to Hannih Jawhariya, the widow of Ibrahim al-Qirra, from a well-known Latin Christian family in that town, who already had a son, Bishara, from her first marriage. By 1820, Nasir was established in Nablus and had two sons, Ibrahim, born around 1817, and Salih, born around 1819. He died probably in 1857. Around ten years earlier his two sons had moved to al-Salt where they acquired some wealth and social standing. Salih was a member of the *Majlis Da'awa* (Juridical Council) appointed in al-Salt immediately after a qa'immaqam was instated in the city in 1285 H.[4] Later, he also became a member of the Majlis Idara and continued to be a very prominent member of the city councils until his death around 1897.[5] Ibrahim, who died around 1890, seems also to have held similar posts; the Salname of 1298 H/1881 AD mentions him as a member of the Court of First Instance, while his eldest son Salīm is also mentioned as a member of the council in 1886. By that time Salīm, who was deeply interested in agriculture – as was his uncle Salih – was already actively developing his own estate.[6]

After their arrival in al-Salt from Nablus, the two brothers became engaged in trading with al-'Awazim tribe, whose shaykh, Abu Wandi,[7] was a member of the 'Adwan confederation. Al-'Awazim formed one group with al-Ghunaymat tribe, mentioned favourably by Burckhardt in 1810, and their dira was Zarqa' Ma' in south-west of Madaba.[8] Evidently there came a time when this trade no longer suited the ambitions of the two established Abujabers, and so they began to deal as well with the 'Adwan when they became involved in the qilw trade. This trade, until the 1860s, was a very important source of income for the nomads in al-Balqa', the inhabitants of al-Salt and the 'Adwan alike. It was essential for the soap-manufacturing industry in Nablus, which bought over 3,000 camel loads annually through the merchants in al-Salt.[9] As a result of the industrial development of caustic soda, a cleaner and cheaper substitute, trade in soda ash became no longer viable. The price of caustic soda continued to drop as the chemical process improved, and in 1866 it went down to £4 per ton from £13.[10] This made it much more economical for the soap manufacturers to use the new raw material as they were buying the soda ash at half a crown (2.5 shillings) the English hundredweight, and also paying 9 piastres in tax to the 'Adwan, another 1 piastre to the people of Salt and transport costs, say a total of £30 per ton.[11]

The changing economic conditions were probably the main reason for the Abujabers' change of occupation just before 1860. Ibrahim and Salih used to travel often to Jerusalem and Nablus where they had relatives, but Salih also went frequently to Damascus where he became friendly with many families, especially merchants and landowners, who belonged to the upper-class and more influential circles in the city during that period. They and other families of their class already owned and were financing the cultivation of lands in the Hawran and this drive on their part must have been strongly promoted by the promulgation of the Ottoman land code of 1274 H/1858 AD.

It may be that Salih was encouraged to cultivate land by the government circles in Damascus who, wishing to collect more in taxes, tried to use their title and control of the land as instruments for stimulating production.[12] The two Abujaber brothers saw the possibility of a new way of life and started their agricultural venture at al-Yaduda, which came to be known as Khirba Abujaber.[13] Often it was simply referred to as Abujaber, following the custom of sometimes calling the geographical location by the name of the clan occupying it.[14] Another similar and older example is Qarya al-'Inab, half way between Jerusalem and Jaffa, better known as Abu Ghush, after the famous family who lived there.[15]

Al-Yaduda, a site 12 kilometres south of Amman 'est une petite localité depuis longtemps en ruines située à peu près à égale distance entre Madaba et Amman, dans le grand plateau de Belqa'.[16] It was interesting to find its name and position rightly placed on a map prepared and published in 1843 by W. Hughes, professor of geography at the College for Civil Engineers in London. South-eastern Bilad al-Sham appears on this map, divided into three parts, and the largest number of settlements appears in the middle section. Seven villages each are shown for the northern and southern parts, while 15 are shown for the middle, al-Balqa'. This is quite an improvement in population dispersion for this part of the country since the days when the Swiss traveller John Lewis Burckhardt visited it in 1812, and wrote only of al-Salt as a settled place.[17]

Al-Yaduda's inclusion in this map, however, is not evidence beyond doubt that it was a populated village in 1843. The first record of this status is a mention in the Salname Vilayet-i Suriya for 1288 H/1871 AD where it is listed among the villages of al-Salt province and its population is given as five *khanat* (households). This list clearly shows that, according to population, al-Yaduda was the smallest of 12 villages in Qada' al-Salt.[18] However, since information during that period was given to the authorities in a form that was planned to suit the purposes and interests of the taxpayer, the Abujabers may have ensured that the responsible authorities at al-Salt gave as the size of their farm the lowest possible population figure to the authorities in Damascus. This would certainly have influenced to their favour the assessment of taxes which the village had to pay to the government every year on its crops. Another reason for keeping the number at a low level may also be attributed to the general wish not to give information about the bachelor *murabi'iya* (farm-hands) who were generally from the Jabal Nablus villages. Most of these men were runaways from the Ottoman conscription system in the better controlled areas of the Empire. They were avoiding the draft which forced them to join Ottoman garrisons and combat forces in far and distant places, such as Bulgaria in the north and Asir and Yemen in the south.

Apparently Ibrahim and Salih had, prior to 1860, developed very good relations with the shaykhs of the Bani Sakhr tribe, probably through their trade in qilw and their hospitality in al-Salt. They were very friendly with a renowned shaykh among them, Rumayh Abu Junayb al-Fayiz, born around 1820 and cousin of Fandi al-Fayiz, shaykh al-mashayikh of all the Bani Sakhr confederation. Rumayh's father Sulayman was the protector of Layard during his short visit to Transjordan in March 1840.[19]

Rumayh, a famous horseman like his father, had probably claimed the tribal rights to the three khirbas of al-Lubban, al-Tunayb and al-Yaduda a few years earlier. In spite of his valour and horsemanship, his control was contested by the Bani Sakhr clan of Khadir, who laid claim to al-Lubban and were far more numerous in number than Rumayh's encampment. They attacked him and after a battle held him captive, probably in 1858. The war-cry, *al-sayih*, went out to all al-Fayiz clans who, as the princely families in the tribe, had to keep control, and who were moreover under an obligation to defend their cousin at a time of need. Al-Fayiz and all their supporters attacked the Khadir at al-Lubban and harassed them to such an extent that the major body of the clan migrated northwards, where they settled in the two villages of Fu'ara and Dawqara to the north-west of Irbid.[20] Al-Lubban was later acquired by Bakhit, another clan of al-Fayiz, some of whom were nephews of Rumayh, and it has remained the property of their descendants until this day. Al-Yaduda was taken, by agreement, by the Abujabers, while a good part of al-Tunayb is still held by descendants of Rumayh himself and his kin.[21]

The Abujabers, well versed in knowledge of tribal affairs, knew very well that Rumayh, famous for his valour, horsemanship and hospitality, would be a good partner and friend. His name was among the first few names of leading horsemen of the Bani Sakhr mentioned to Nazim Pasha, the wali of Damascus at the end of the nineteenth century.[22] His raids against the Ruwala tribe inspired one of their famous poets, Khalaf al-Udhun al-Sha'lan, to write of him in a well-known *qasida* (poem) commemorating a bloody battle between the Bani Sakhr and the Ruwala, in which he proved to be the major figure. A free translation of part of the poem may help to give some idea of his position and importance as the horseman of his tribe. It is to be remembered that the poet, while mentioning Rumayh by name, was actually addressing the whole enemy tribe, the Bani Sakhr:

> Oh, Rumayh, had it not been for your treachery,
> You are not really bad,
> You persist well in battle and your hospitality
> to all is renowned,
> The shaykhs of noble tribes are not
> subordinates to follow behind,
> They too have proud noses and in their domains
> good mounts abound.[23]

Rumayh was a wise man who wanted to establish once and for all his right to the lands that he had laid claim to, especially after his encounter with the Khadir. The Abujabers were ready for a new venture to satisfy their ambitions. Many of the factors for a successful partnership were therefore present. The original agreement, concluded between the two parties around 1860, stipulated that Ibrahim and Salih would till the land, pay all expenses and, after harvest, Rumayh would receive half the crop and retain half the cultivated land, while the Abujabers would receive the other half of the crop and acquire the other half of the land.[24] The expression used for this agricultural venture was *futuh al-ard* (the opening of the land); for the actual tilling of the fields, the expression used was *kasr al-ard* (the breaking of the soil). These expressions, strange as they may seem, were obviously used because the lands in that district had not been tilled before, as far as human memory could recall. Possibly the whole area was transformed, through neglect, into pasture land during the fourteenth and fifteenth centuries when Bilad al-Sham was smitten by devastating outbreaks of plague.[25] The records of the Daftar Mufassal al-Jadid compiled in 1005 H/1595 AD do not mention al-Tunayb or al-Yaduda among the inhabited and tax-paying villages.[26] The most eastern area of cultivated land in al-Balqa' was that of al-Salt, some 30 kilometres to the north-west. It is therefore not surprising that the venture, difficult as it seemed, carried with it much promise of success in virgin lands. The tilling itself was a difficult task, since the soil was so hard, and added to this the countryside was also covered with bushes, plants and roots. Outstanding among these were varieties like quram, bilan, and injil which had sometimes to be removed by pick and shovel. Ploughing was carried out by pairs of strong oxen (faddans) that were generally purchased from villagers in the Jawlan and Hawran at the initial stages.[27] Later, the development of animal husbandry permitted the raising of a strong strain of oxen on the estate itself and mules were used in larger numbers. The ploughshares, made of iron in the old-fashioned blacksmith workshops in al-Salt, were heavier than the usual standards and therefore went deeper into the soil.[28]

The agreement, entailing a new form of partnership among people engaged in agriculture, presented another interesting aspect of great importance. Contrary to prevailing conditions, it left no room for any element of khawa (protection). Naturally Rumayh was expected to protect the crops from his own tribesmen, the Bani Sakhr, but Ibrahim and Salih seem to have become so well established as dignitaries among their own people in al-Salt, their native town by this time, that they may not have even considered the possibility. Their position must have been strengthened by two facts. Firstly, as leading figures, they were entrepreneurs who never handled farming with their own hands but hired people to do it for them.[29] Secondly, they had a great love for lavish hospitality which they offered in their homes in al-Salt and which they continued in the new settlement. Such generosity is a cause of reverence and great endearment among all Arabs, whether they be nomads, semi-settled or urban. Beduins spoke in prose and poetry of the Abujabers' hospitality, while people in al-Salt appealed to the deity to send the rains and irrigate the Abujaber fields during the traditional parade in drought-stricken years. Their plea was simple and to the point:

> Mother of rain, you eternal,
> Please irrigate our thirsty fields,
> Please irrigate the crops of Abujaber
> whose hospitality is continuous.[30]

Although the original venture was started by Salih and his older brother Ibrahim, the real force behind the project seems to have been Salih. Family chronicles relate that he was a very active man and an aggressive entrepreneur whose ambition was boundless.[31] Towards 1884 his two elder sons, Farhan and Frayh, had already for many years been helping manage the estate. Ibrahim's two elder sons, Salīm and Sālim were likewise engaged in its management. The usual differences occurred and in 1885 Salīm went to Jerusalem where he borrowed from the Ottoman Bank a few hundred gold French pounds.[32] He used the sum to buy Quraya Nafi', a farm of 5,454 dunums, and a similar area in al-Juwayda from the Balqawiya tribe. He settled here with his brothers and resumed his agricultural activity in close co-operation with the shaykhs of the Haddid clan who were very good friends of his father. A year later, either he was pressed by the bank for the settlement of the loan, or he preferred to save the interest (unfortunately we do not have a record of the rate at which it was charged, but it could have been high, in line with the risks involved and the ruling rates then). He therefore arranged that his wife Helena, a Greek lady whose family was residing in Jerusalem, should sell her share in a house at the Damascus Gate for 625 French gold pounds, of which 600 were paid back to the bank.[33]

Obviously the relations between the two branches of the Abujaber family were no longer as close as they had been, but the elders always maintained that they were cordial and friendly.[34] The two growing clans were known from 1890 as Jawabra al-Yaduda and Jawabra al-Juwayda. A third group,

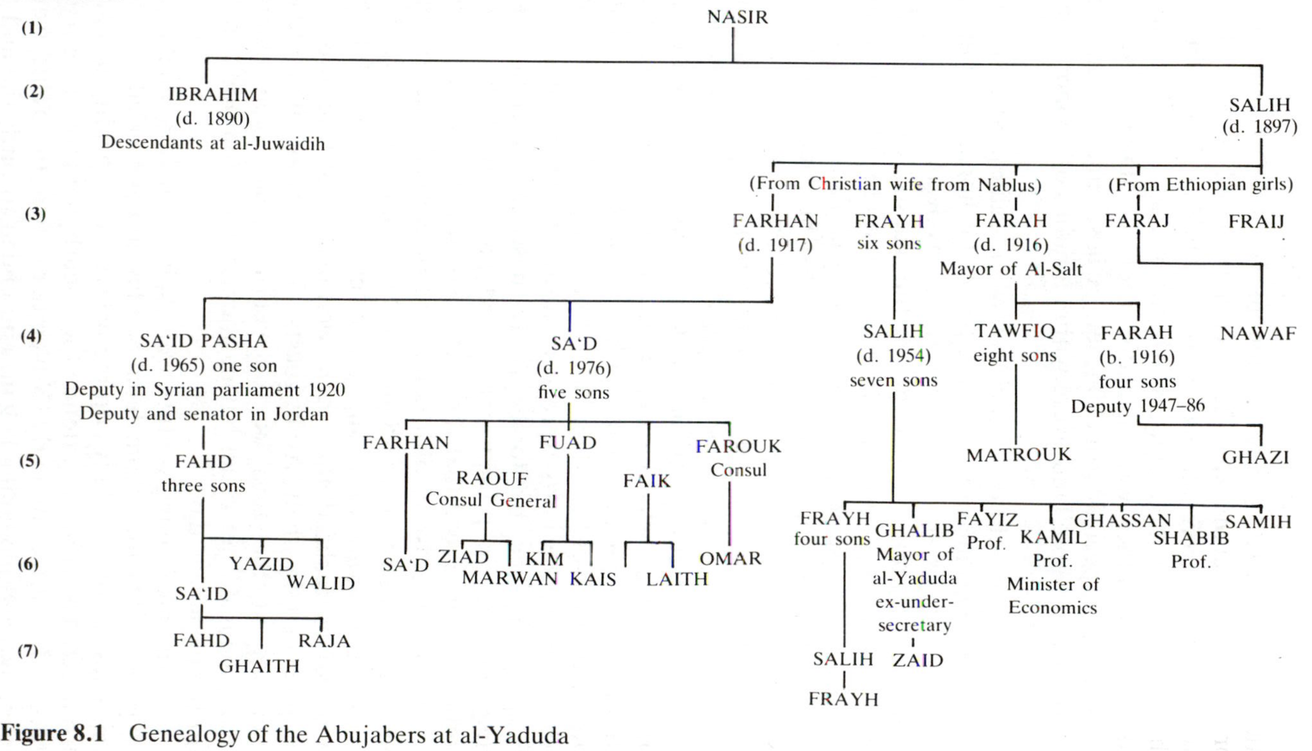

Figure 8.1 Genealogy of the Abujabers at al-Yaduda

the Circassians who settled in Amman after 1878, distinguished them in a different way. They called the Abujabers of al-Yaduda, who were all big in size, Abujaber *Ishwa* (big in Circassian) and their cousins Abujaber *Tzig* (small). These three communities were friendly after the initial stage, as would be expected from farmers whose major aim was stability and whose lands were within a radius of no more than 15 kilometres.

The story is told of an incident when Salih paid a visit to the Circassians in Amman with a few of his horsemen. One of his companions suspected that the hosts were preparing to slaughter a horse in honour of the guests, following the custom in the Caucasus. He immediately told Salih. As Arabs do not eat horse meat, Salih tried to explain to his hosts, delicately so as not to give offence, that he could not accept such hospitality; but the hosts insisted that since he was an important guest they had to honour him in proper form. The language barrier was of course great; the Circassians in their broken Arabic could not convey their point of view, nor could they understand why this important guest was declining their traditional hospitality. After an hour of anxious debate, Salih decided to leave; but when he and his party made a move for their horses, the hosts were very insulted and threatened to stop them even by force. A Damascene merchant who happened to be present intervened and through his good offices it was agreed that the guests would dine instead on the celebrated Circassian dish composed of cooked chicken and boiled, crushed wheat.[35] Salih and his companions were thus saved the agony of eating horse meat that day and the Circassians realized that their well-meant gesture was not appreciated by their Arab guests. The custom died rather abruptly at the turn of the century, and even elderly Circassians do not recall having themselves eaten horse meat in their younger days.[36]

To run his farming venture well, Salih brought a number of experienced farmers from Jabal Nablus and employed them as his harathin or murabi'iya on the basis of an agricultural year's contract. The contract extended for nearly ten months of hard work. It started in early November when the rains were expected and ploughing began, and ended towards the end of

August when al-baydar (threshing) came to an end and all the crops were safely in store. The harathin were mainly from the villages of ‘Awarta, Awdala, Agraba, Bayta, Qasra and Sarra in Jabal Nablus, but there were also some from Jabal al-Quds (Jerusalem area), Ramallah, al-Salt and sometimes from places as far north as Hawran. In this connection, I was told by Lutfi Muhammad Hasan al-Khatib (Abu Isma‘il), in 1970, that people of his village, al-Qibab in the district of Ramla, Palestine, had a proverb that was popular among them before 1948: *Bayt ‘Inan wa-Bayt Illu, in kithru wa-in qallu harathun ‘ala Abu Jabir* (*Bayt ‘Inan* and *Bayt Illu*, whether increasing or decreasing in number, are workmen at Abujaber).[37]

Al-Yaduda lay on the border between the two large tribal confederations of al-Balqa’ (Ibn ‘Adwan) and Bani Sakhr. It became neutral territory under Salih, who was a Salti himself and therefore of the former confederation, and for many years was the meeting place of the shaykhs of the two alliances. It had eight khirab as neighbours and although al-Tayba – a village that mushroomed in the lands of Khrayba al-Suq after the 1967 war – may now claim over 25,000 inhabitants, until 1960 al-Yaduda had the largest population. Apart from Tunayb, which has a total area of 25,607 dunums, al-Yaduda was also the largest in area with its 22,087 dunums and over 500 extra dunums owned by the Abujabers in neighbouring Umm al-Hanafish (now known as Umm al-Basatin).[38] As far as cereal production was concerned, al-Yaduda was by far the largest in all al-Balqa’ and probably in all south-eastern Bilad al-Sham. Its fields were famous for their ability to give a reasonable crop even if the year’s rainfall was well below the average. Its situation is outlined in map 8.1, and table 8.1 where the neighbouring villages have been enumerated clockwise, starting with al-Lubban in the east. Prior to the 1850s, all of the three khirab in the Bani Sakhr domain, together with al-Yaduda, were deserted. Beduin encampments were found in them during the spring and summer months only. In the other five khirab, on the other hand, members of the Balqawiya clans were permanently living all the year round. Some were in encampments, but others preferred to avoid the extremely cold weather in winter by living in relatively more comfortable accommodation in caverns and caves.

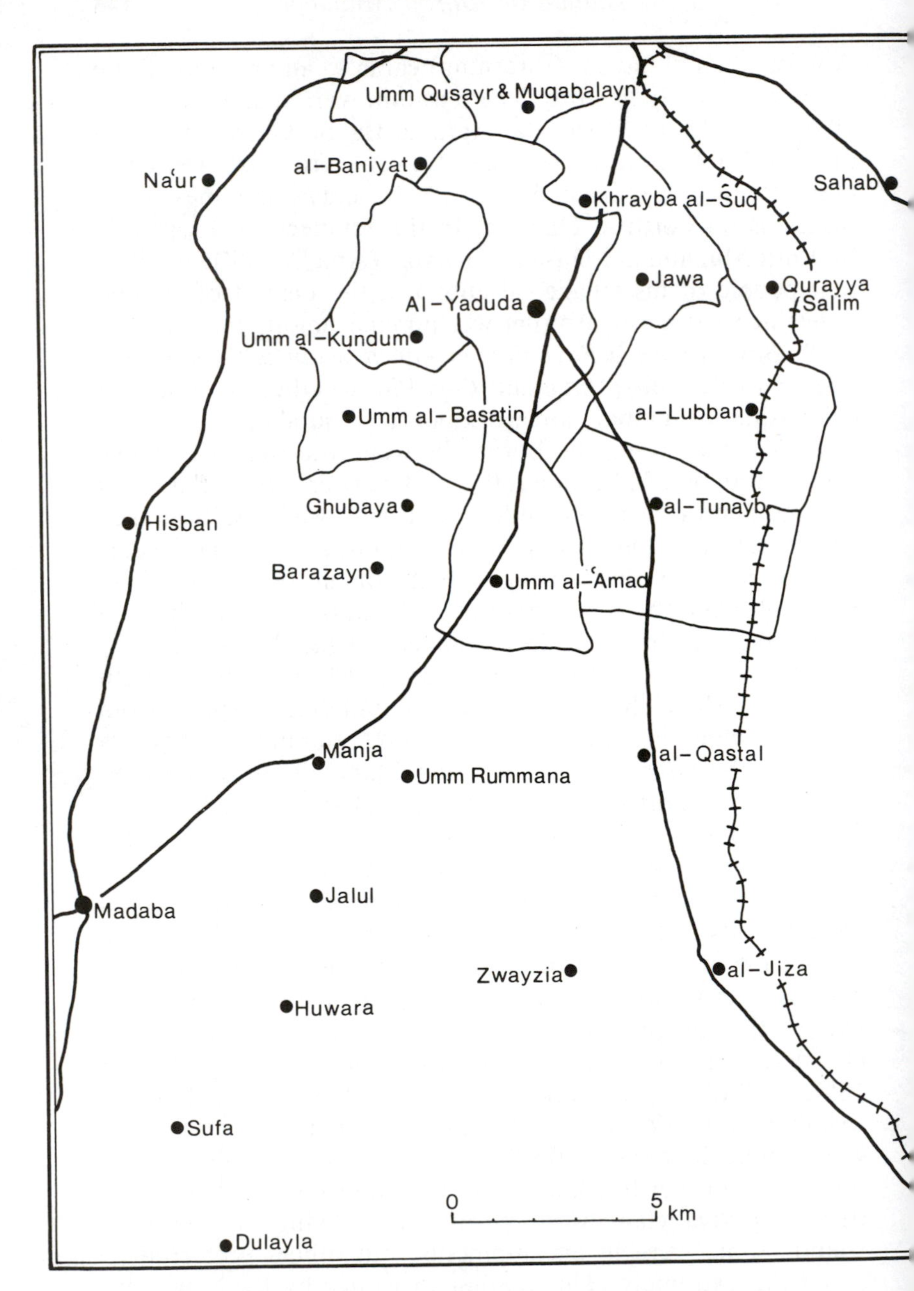

Map 8.1 Al-Yaduda and its neighbouring villages

Table 8.1 Al-Yaduda and neighbouring villages

Kirba	Area in Dunums	No. of plots	Clan	Owners tribes	Confederation
Al-Lubban	21,165	87	al-Bakhit	al-Fayiz	Bani Sakhr
Al-Tunayb	25,607	21	Abu Junayb (al-Junaybat)	al-Fayiz	Bani Sakhr
Umm al-'Amad	16,379	14	al-Sattam but mainly Shaykh Mithqal Ibn Fayiz	al-Fayiz	Bani Sakhr
Umm al-Hanafish[a]	11,573	268	al-Shahwan & al-Yisifa	al-'Ajarma	al-Balqa'
Umm al-Kundum[b]	8,904	29	al-Shahwan & al-Yisifa	al-'Ajarma	al-Balqa'
Al-Baniyat	5,422	294	al-Mari'i	al-'Ajarma	al-Balqa'
Umm Qusayr and Muqabalayn	6,795	128	al-Khusaylat of al-Da'ja	Balqawiya	al-Balqa'
Khrayba al-Suq and Jawa	13,796	304	al-Zufifa of al-Haddid	Balqawiya	al-Balqa'
Al-Yaduda	22,087	88	Salih and sons	Abujaber of al-Salt	Bani Sakhr[c]

[a] The name was changed in the 1960s to Umm-al-Basatin.
[b] Bought by al-Bisharats of Nablus, cousins of Abujabers.
[c] Although technically neutral, it was considered to be in the Bani Sakhr domain.

The figures for areas have been derived from the records of the governmental survey which was carried out by the Lands and Surveys Department of the Emirate of Transjordan. The work was undertaken by teams of supervisors and surveyors directed by land-settlement judges who measured land and settled disputes in every village between the years 1935 and 1942. In accordance with the laws prevailing then, a proportion of the land of every village, usually barren or rocky land that was unfit for cultivation, was set aside and registered in the name of the government for afforestation purposes. In the case of al-Yaduda, the area was 1,114.753 dunums or nearly 5 per cent of the total area. Forty-five years later, these lands are fully grown pine forests, certainly a great asset for the countryside and its population.

A very ambitious man, Salih was naturally interested in a large-scale operation and did not therefore adhere to the methods already prevailing in his town and the surrounding countryside. The muraba'a system (sharing of crop whereby labour receives a quarter) was still, however, the basis for his venture. This was a system whereby:

> rich individuals who have lands maintain the husbandmen in their own houses on the farm and, in addition to their food, give them one-fourth of the produce of the soil to be divided equally among them, reserving the other three-fourths for the landlord or occupier. This ratio of division is always observed whether the produce of the farm consists of corn, fruit, and oil raised from the land or cattle born on the soil since the commencement of the husbandmen's servitude.[39]

Salih wished to develop his system in his own way. He began by separating the farm's household, which became known as the *baylik*,[40] from his family house which was nearby and accommodated the nearest members of the family.[41] Within a few years his venture developed into a very large farm and became known as a *sawama*, a word applied to no other estate in the whole area. It probably derives from the Arabic word *sa'ima* meaning animals in general.[42] Evidently it was applied to large farms which, together with many fields, owned large numbers of animals of different kinds for work, milking and breeding.

The development at al-Yaduda was gradual but continuous. Salih probably started with ten or twelve faddans in the mid-1850s and the number grew over the following few years. The records of the Latin patriarchate of Jerusalem described Salih in 1868 as the 'chef d'une famille puissant qui avait de grands biens à Yadoudah, un homme intelligent mais trop infatué de lui-même'.[43] It is safe to assume that during the years 1865–8 he was already running 50 or 60 faddans. The Ottoman government had no physical presence in al-Salt or anywhere in al-Balqa' until 1867: Salih's success was brought about by his friendly and good relations with his neighbours and without intervention on his behalf by the government. In the year 1879, the traveller Laurence Oliphant, who advocated the settlement of newcomers in Palestine, mentioned that while in al-Salt he heard that Salih Abujaber was cultivating an area of 60 faddans. Oliphant was unable to see why a Protestant

Syrian could succeed in such a venture when other settlers could not.[44]

Unfortunately there are no records now of the way in which the estate was run but the information provided by Muhammad al-Rashid who was born at al-Yaduda in 1896 and worked there continuously until his retirement in 1964 is most helpful. He was told by his father that around 1890 Salih Abujaber, who was already old, divided the estate among his three sons. Each had a rabta with the oldest being given an extra rabta as a *kabra* (additional share as a sign of reverence and seniority) comprising six faddans. The farm was thus divided as follows: Farhan, the oldest son, ran two rabtas of 15 faddans each; Frayih, the second son, and Farah, the third son, each ran one rabta of 24 fadans. Farhan and Frayh supervised the farm while Farah became Mayor of al-Salt and stayed in the post until his death in 1916. His two older brothers died during the typhus epidemic that swept through the countryside in 1917.

Besides his three sons, Salih had three daughters, all of whom were married when he divided the land. Following the custom of those days, the daughters were not given anything. It was told in family stories that when Salih was on his deathbed in 1897, his wife, Samra, asked whether he had willed anything to the daughters. Salih angrily replied that his sons were good brothers to their sisters and would undoubtedly bring an embroidered jacket for each. (These expensive jackets seem to have been very fashionable at that time and were specially ordered from Damascus.) The matter rested there and the three daughters had no share in the legacy.

In addition to the 78 faddans run directly and for their own account by the three sons, there were also 18 faddans run on a partnership basis by the fallahin who had their own murabiʻiya.

		faddans
Rashid al-Badawi	from ʻAwarta in Jabal Nablus	2
Irshayd al-Badawi	from ʻAwarta in Jabal Nablus	2
Muhammad al-Salih	from ʻAyn Sinia in Jabal Nablus	2
Muhammad Abu Awda	from Sinjil (St Giles of crusader times)	2
Shihada al-ʻId	from ʻAwarta in Jabal Nablus	2
Al-Sawara brothers	from a village in Hawran	8
Total		18

These fallahin were a different group from the harathin. With a few others they were considered the dignitaries of the village and as such were given special assistance and facilities. Generally they paid to the landlord only 40 per cent of their crop and landlords bore the ‘ushr on that share. All of them came to al-Yaduda in their younger days and because of personal merit developed into the leader class. They rendered continuous services to the estate by carrying out errands in the neighbouring villages and accomplishing missions in the farm itself. They were given mounts whenever needed and their judgement in important matters was accepted by the landlords. Being of farmer stock they usually handled affairs related to farming and agriculture in general, while their colleagues of beduin stock dealt with tribal and beduin concerns. The latter were from different tribes and usually they were respectable members of their own societies. Their leader was Sayyah al-Nimr al-Fayiz, cousin of the paramount shaykhs of the Bani Sakhr. Others were his cousin Arrar, Mahmud al-‘Ajalin of the Zufifa tribe, and Abu Salama al-Minhakit of the ‘Ajarma tribes. This group did not farm any land to generate extra income and the estate had to provide them with practically everything in the way of food and dress.[45] Each of them had a flock of sheep and goats to support their household needs and as an extra source of income.

Another important group in al-Yaduda were the shuyukh al-rabtat and their duties were varied and many. They divided work among the harathin in their rabtat, decided which fields were to be worked, and managed the rabta affairs as a whole. In consultation with the mu‘allimin (landlords) they decided the crops to be planted and the time of seed throwing, a very important matter in years of late or little rainfall. During the 1890s, the four rabtat had shuyukh from the Jabal Nablus area. This composition was in line with the composition of the murabi‘iya and other members of the work force. In later years the estate had workers whose grandfathers had been born on it but they were always known by the birthplace of their grandfathers. Thus practically every person was called after his village, for example al-‘Awartani, al-‘Aqrabani, al-Qasrawi, al-Bitawi, al-Sufani and al-Wahdani. However, since the 'Awarta group was by far the largest and perhaps formed more than 60 per cent of the winter-time labour force of more than 200, it was natural that the shuyukh al-rabtat would be mainly from among them.

Every shaykh rabta was ex-officio the *bathar* (seed thrower) during winter, and the kayyal during summer when the crops were ready on the baydar (threshing ground) and during the rest of the year when quantities were sold from the stores. For their service and extra responsibility every shaykh rabta was given the crop of a *shukara*[46], an area sown with two kayls of seeds (a kayl is twelve sa's and a sa' is six kilograms of wheat), in addition to his share of the rub' (quarter allotted for the labour force). In a good year the share of the rub' would be some 300 sa's of wheat weighing 1,800 kilograms, to which was added the share from barley, lentils, chick-peas and sesame. The additional share of a shaykh rabta over that of a murabi' could in a good year be another 400 sa's of wheat from his shukara.

During the first few years Salih was content to use caves for the stabling of animals and storage of seeds and produce. Some of these were demolished during road building and modern housing projects, but there is still intact a large cave called maghara al-baylik (the baylik cave) which is situated on the top of the hill. It has a few compartments and must have been sufficient for the farm's requirements when Salih was running the first twelve faddans. In subsequent years he built in stages the buildings that were needed. A good supply of stones was available in the ruins all around and the dismantling of these old buildings was so systematic that by the turn of the century the hill of al-Yaduda was rebuilt with houses and four units of bawayik.

In the 1870s Salih and his two older sons ventured on a new course when they discovered that the ordinary type of crude building of the area did not serve their purpose. They therefore brought in masons and builders: from Nablus they brought Nur al-Din 'Abd al-Razzaq and from Ramallah Abu Jiryis Ziyada, both of whom were known as master craftsmen. They started with the guesthouse on the northern side and the unit of stables next to it. To gain an idea of the size of the project, it is well to mention that every one of the stables comprised a unit of 16 *'aqds* (Roman vaults), seven by seven metres each, or a total of 784 square metres in built areas. Over the years four such units were built; their roofs were more than five metres high, so that parts of them were used for the storage of huge quantities of hay (chaff) during the winter months. There was enough ground space for the hundreds of animals that lived on the estate as well as for some of the flocks of sheep and goats that happened to be in the vicinity. The map of the built area in al-Yaduda (map 8.1) will give an impression of its extent.

The bawayik and the courtyards around each of them were the centre of agricultural activity. Each rabta was allocated one of them and in each there was a household wakil who was in charge of the compound including its stores and equipment in use. To assist him in performing his duties he had a bawwab (doorkeeper) who was generally a stranger and often a maghribi (from al-Maghrib).[47] Other assistants were three or four qatariz, two 'izbiyat (women cooks-bakers) and one saqqa who in addition to carrying water also arranged the transport of food to the workmen in the fields. The three establishments had at their disposal the services of a carpenter and an assistant who worked on the repair of the wooden ploughs and the other pieces of equipment. The household had also a jammal and an assistant who looked after the camels and all matters related to them. In winter the household may have had one waqqaf but in spring and summer two or three were needed. Their duty was to visit the fields, sometimes at night, to ascertain that the boundaries with neighbours were not tampered with and that the crops were not molested by thieves or grazing flocks. These men were usually from clans around al-Yaduda and were normally issued with rifles and ammunition. Sometimes they fired in the air to scare off intruding shepherds and alert their colleagues in the village. Usually such a call was answered by two or three riders who rode out to discover what was happening and to assist in the settlement of the affair. The good-neighbour policy was the general rule; it was in their own best interests, the party at fault was shamed and in almost every instance it won the Abujabers and their men new and good friends.

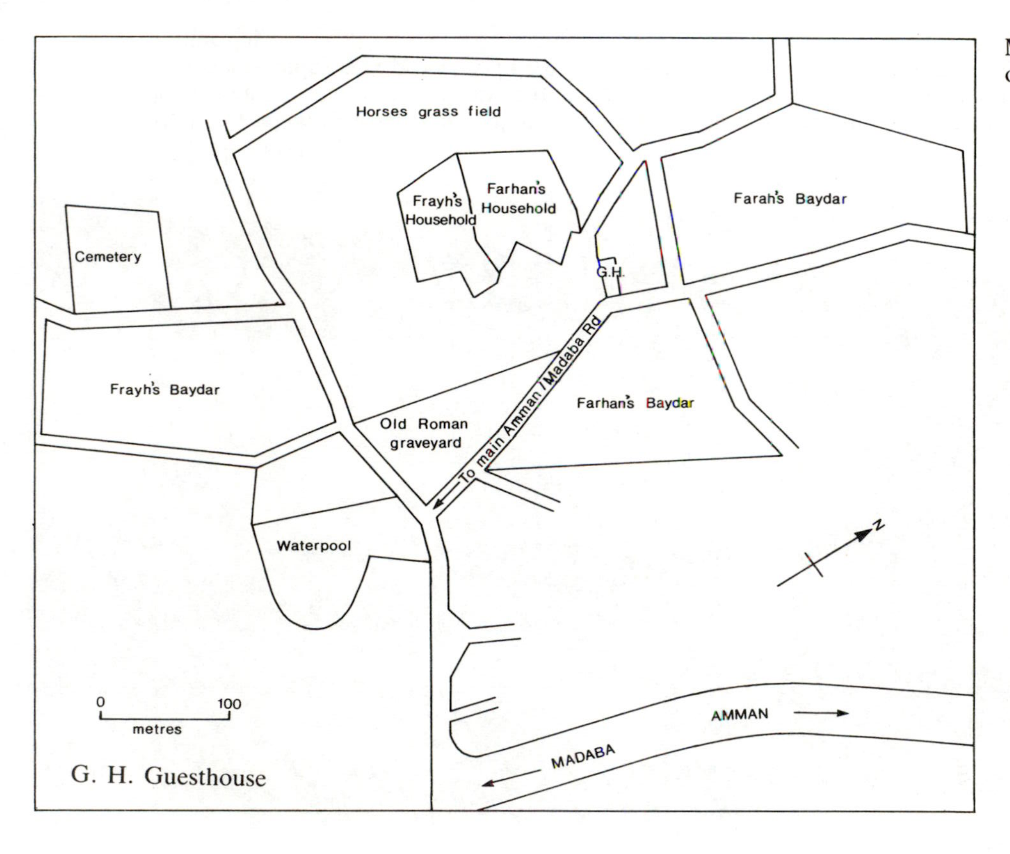

Map 8.2 The village of al-Yaduda, *c*.1900

Al-Yaduda *c.*1900

Probably taken by a Jerusalem photographer at the turn of the century, it shows the three-storey family living quarters that were destroyed by an earthquake in 1927. The horseman on the white mare was an Abujaber landowner, probably Frayh. The men next to him are workmen, heavily dressed for a cold winter day.

Al-Yaduda, seen from the north (1985)

Hospitality and generosity played an important role in the life of al-Yaduda and the welfare and prosperity of its owners and those who lived there. Salih and later his sons had an open house and one of the first buildings to be erected was al-madafa. It appears in the foreground of the photograph which was probably taken just before 1900 by an unknown photographer. It shows one of the mu‘allimin on horseback standing in front of a line of some of the workmen. Al-madafa was a simple but sturdy construction composed of an entrance hall four by nine metres, flanked on either side by two large rooms of nine by nine metres each. Just behind the main gate, there was a big niche in the thick wall which contained a large *jarra* (jar) that was continuously filled with fresh water for drinking by guests as well as by people of the village. The water was drawn from the jarra by a long-handled copper *tassa* (mug) which people drank from. Guests went from the entrance hall to the eastern room on the right which had east-facing windows overlooking the baydar and north-facing windows. In the middle of this room there was a *nuqra* (fire hearth) 1.5 metres in diameter which served for coffee-making as well as heating during the cold winter months. The best available firewood was burnt so as to lessen the flow of smoke and if the guests were really important, the fire was lit outside in *manqals* (fire containers) which were brought in only after there was no more smoke. Coal bought from the ‘Ajlun area was sometimes used but generally was reserved for use in the city house at al-Salt.[48]

Al-madafa always had a regular attendant, called the *qahwaji* (coffee maker), who was helped by a young assistant. When guests came in, the qahwaji immediately welcomed them and started laying carpets and quilts, of which he had a sufficient quantity, on the ground, next to the walls. He placed enough of these to accommodate the number of guests on hand. The room, having only one door and four windows, that were in fact openings or niches in the walls on the eastern and northern sides, could accommodate around 40 men seated with their backs to the walls and windows. If the number of guests was bigger, the qahwaji would usher some of them to the western room where they would be seated likewise. It was then the duty of the young attendant to run uphill and inform

Salih or one of his sons about the arrival of their guests. Generally their identity, if they were not from the immediate neighbourhood, would not be known since it is against Arab tradition to ask. However, the qahwaji, having acquired experience over the years, was able to guess the guests' tribe and social status by their looks, dress, dialect, by the number of attendants in the party and the quality of the mounts. By the time the host arrived to welcome the guests, the qahwaji would already be roasting the coffee beans (*'adaniya*, imported from Yemen via Aden) to prepare *al-qahwa al-sada* (black coffee) to which he later added, in generous doses, *habb al-hal* (cardamom seeds). This spice was a very important element in the coffee-making ritual and its use in good quantity was considered a sign of the importance of the guests and the generosity of the hosts.

After warmly welcoming the guests, the hosts would ask about their welfare and that of the tribe and their flocks or herds depending on whether they were camel-breeders or sheep-breeders. *Kayf halkum*? (How are things with you?) *Kayf intum*? (How are you?) would be said a few times, and expressions like *In sha' allah mabsutin* (May God wish that you are well) and *Ahlan wa sahlan* (welcome, literally, You have come to your own people and found ease) would be uttered every now and then. If the guests were old friends, the conversation would take a more personal turn and questions would be asked more specifically about brothers, cousins, pastures, crops, herds and flocks.

Depending on the time of arrival of the guests, the host would issue instructions for the preparation of the coming meal, often in the form of the traditional *mansaf* (a large platter carried by two or three men on which thin bread is laid heaped with rice, with meat on top surmounted by the cooked head of a lamb). In most cases rice was used but sometimes in summer, and only after consultation with the guests, *farik* was cooked instead. This is crushed wheat that has been burnt or roasted when green in June; it was considered a delicacy but was not traditional. One lamb or more was killed depending on the number of the guests, the average being a lamb for every ten or fifteen men. The meat was boiled in yoghurt and the rice cooked with samna (Syrian butter ghee). Lunch was usually offered around 2.00 p.m. and dinner between 6.00 and 8.00 p.m., depending on the time of year; usually it was served at an earlier time in winter and a later time in summer. Breakfast was composed of fried eggs, when available, laban or *labna* (yoghurt or yoghurt that has been strained through a cloth bag), dibs (syrup made from boiling the juice of grapes or dates), olives and *halawa* (sweetmeat made from sesame oil and sugar). A speciality was a dish called *luzayqiyat* made from *khubz shrak* (thin bread) covered when very hot with butter and granulated sugar.

Although al-Yaduda had an abundant supply of lambs and kids from its own flocks, additional numbers were bought in the season by the hundred. These were killed for food almost every day, but not every guest was entertained to a mansaf. This was perhaps because cooking such a dish required plenty of extra preparation and work. First a lahham (butcher who knew how to kill and skin a lamb) was needed. Then an 'izbiya had to be brought in to make the khubz shrak, which had to be fresh (in contrast to khubz tabun, bread cooked in the traditional oven), as well as cook the rice. So if the guests were workers or ordinary tribesmen they were offered breakfast dishes and the qahwaji would most probably be their companion at the table. With other more important guests the traditions prevailed and the hosts did not partake of the food offered to the guests. It was their solemn duty to see that the guests were comfortable and being well served. Their answer, whenever asked by the head guest kindly to participate, would be *al-mu'azib rabbah* (the host gains the pleasure from his guests' comfort).

In spite of the fact that al-Yaduda was adjacent to the khirab of Bani Sakhr, their system of killing a *jazur* (a she-camel specially fattened) was not adopted. Its people preferred to offer lamb meat to their guests. Camels and oxen were killed but mainly on feast days and at the start and end of harvest. It was also done on special occasions like the birth of an Abujaber son or when the landlord was cured from illness. Everybody on the estate was invited to join the celebrations and participate in the meal that followed. On these occasions rice perhaps was too expensive and in any case had to be bought in; therefore they used *jarisha* (crushed

حكومة الشرق العربي
رئاسة النظـار
عدد: حضرة الفاضل سعيد باشا ابو جابر المحترم
٤٩٦٠ / ١ / ٢٥/٥

بالاشارة للاجتماع الذي عقد نهار امس بحضور سمو امير البلاد المعظم
ووجود كل من المستشار المالي المستر كركبرايد والسكرتير العام عارف بك العارف والمستشار
الافرنسي المسيو بونه وقائد المتطوعة مسيو كبرناك والشيخ خلف الكليب السردي ومشايخ بني صخر .
أود ان أؤيد فيما يلي ~~الارقام التي سبق ذكرها~~ حول الاغنام التي نهبها الشيخ خلف المومأ اليه
قبل تسعة شهور واليكها :—

عدد		عدد	
٢٠٠	الاغنام المنهوبة (كبيره)	٩٦	عبر وخرفان
٣٠	الى الراعي	١٤	للراعي
١٧٠		٨٢	
٧٧	واصل ليده	٢٢	واصل ليده
٩٣		٦٠	
٣١	المأخوذ من قبل اهالي الجبل الدروز		
٦٢			

وقد اقسم الشيخ خلف المومأ اليه على ان هذه المواشي لم تدخل لذمته ثم افاد ان
قد وصله منها (١٦) راسا كهدية وهي التي تم القرار على ارجاعها الحكم كما انه تعهد
بارجاع الواحد وثلاثين راس التي اخذها الدروز من اهل الجبل بمساعدة الحكومة واظهر المسيو
كرناك قائد المتطوعة رغبته في المساعدة في هذا الامر .

والسلام عليكم

٩٢٧/٨/٢٩ —

رئيس النظار

The return of 296 sheep by a marauding party to the Abujabers wheat), which was boiled in water with some butter or samna and served with the meat. The name used for the crushed wheat dish is ‘aysh.

Guests normally stayed overnight and left early next morning, but some stayed longer and were continuously given the same courteous attention. The age-old sacred Arab tradition of offering hospitality for three and one-third days without asking about the identity of the guest or the purpose of his visit was strictly adhered to. However, this sometimes resulted in peculiar incidents. Salih was once in Damascus where he bought from a merchant a quantity of ropes and strings for the amount of five Ottoman gold pounds. Since he was short of ready cash, the merchant kindly agreed to sell on credit and come to al-Yaduda to collect the debt sometime in August during al-baydar. When the man called, Salih had forgotten about him and the debt. He received him well and offered a mansaf for every lunch and dinner over three days. The Damascene merchant was too shy to ask for the money and on the fourth day left the guest-house for home empty-handed. On being asked why he did not collect the debt he replied, ‘I was not going to be less generous than him. Anyway I have eaten the five gold pounds in mansaf lambs.’ Salih, it is said, heard the story later and settled the debt during his next visit to Damascus.

Al-Yaduda and its owners earned the reputation of being the most hospitable house. It was favourably mentioned in beduin encampments all over Badiya al-Sham. This was a great advantage for Salih, his sons and their farming operations. They were very well respected and tribal shaykhs were ashamed to have differences with them for material gain. The property, in general, was immune from molestation or plunder. They were given friendship in return for their hospitality. As late as 1928, this attitude prevailed when, in the presence of Amir Abd Allah, the prime minister of Trans-jordan presided over the meeting of Jordanian, British and French high officials to sort out matters concerning the 296 head of sheep belonging to the Abujabers and plundered by the Sardya tribe and the Druzes. The document gives the story, but all the 296 head were returned within one month of the meeting.

The success of the enterprise of Salih Abujaber at al-Yaduda was noticed by townspeople in al-Salt and especially by two groups of his relatives. Towards 1875, his three nephews, Ibrahim, Saliba and Salti, sons of Bishara al-Qirra, came from Nablus and settled in al-Salt where they became known as al-Bisharat.[49] He helped them to acquire Umm al-Kundum to the west of his estate and their descendants remain the owners and farmers of this village. Salih's in-laws, the Qa'awara,[50] thereafter acquired half of Qurayya Salim to the north-east. The Nabulsi family,[51] who had come to al-Salt from Nablus a few years earlier, acquired a good part of the village of Hisban. Another pioneer who migrated to al-Salt from Damascus, Khayr Abu Qura,[52] acquired al-Rajib, and his descendants, now known as the Khayr family, continue to own the lands of this growing village. Al-Sharabi bought lands in al-Baniyat,[53] while Salim al-Fayyadh[54] acquired many fields in Jawa and Khrayba al-Suq. People were going out into the new lands and agriculture was spreading year by year.

The prosperity of al-Yaduda grew towards the end of the nineteenth century. Before their deaths in 1965 and 1976, Sa'id and Sa'd, the sons of Farhan, used to render vivid pictures, during family gatherings, of life at al-Yaduda when their father and uncles were running it. It was an operation of continuous toil and work but one that was rewarding, not only in economic returns but also in human values. It provided work for at least two hundred families at the time when people needed work most. It provided, too, an atmosphere of complete tolerance and understanding between Christians and Muslims, an atmosphere which spread beyond the precincts of the village and out into the countryside.

The venture was supported, they said, by animals and rudimentary equipment. The following list of estimates provides an idea of the scale of the operation. Numbers increased or decreased within a margin of 10–15 per cent depending on the crops and economic requirements. In bad years, for example, it was not unusual to sell animals, both to improve the cash flow and to save on fodder consumption. Generally sheep, goats and cattle were the most current or movable, and it was common to sell a flock of 200 ewes or 200 goats when cash was needed.

Animals at al-Yaduda towards 1900

1 150 plough oxen, sometimes bought from the Jawlan but generally raised on the estate. This flock was called *al-'ammal* or worker.
2 600 cows and calves, in complete distinction from the first, called *al-fadal*, or the extra.
3 25 mules, generally bought from those who produced them through the copulation between a mare and an ass. Following beduin tradition the Abujabers did not allow it to happen to their mares.
4 50 horses and mares.
5 250 donkeys and she-donkeys for breeding.
6 60 load camels. Herds of she-camels were never kept at the estate, because they were often plundered by raiding parties.
7 3 dancing horses called in Arabic *rahwan* (pl. *rahawin*). These were especially fast riding horses, trained in Damascus. They fetched high prices in comparison to ordinary horses.
8 1,000 goats.
9 3,000 sheep.
10 There were always a number of hens for egg and chicken production on a medium scale, when chicken pestilence permitted. Pigs, however, were never raised or allowed on the estate. They were considered, following Islamic tradition, *najis* (dirty or unclean).

The agricultural economics of the farm were greatly dependent on the rainfall in every year. Al-Yaduda had the advantage of being within the area that gave a crop even in the bad rainfall years, an area generally considered to be west of Khatt Shabib, an imaginary line named after a local ruler of Transjordan in the tenth century, which traverses the country from north to south and mainly coincides with the Hijaz railway. Al-Yaduda is a few kilometres west of the railway, and its fields, in 120 years of cultivation, have never failed to give a crop. The yield, though, has varied between 6 and 20 to the measure. An annual yield of 10 to the measure has been assumed for the purpose of computation of al-Yaduda's crops in the tables 8.2 and 8.3. To facilitate the process, kayls of cereals have been transformed into kilograms, figures have been rounded off and the accounts computed per faddan.

Table 8.2 Accounts of large-scale farming at al-Yaduda around 1900 (computed per faddan in one agricultural year)

Item	Wheat	Barley	Others	Total
Fixed expenditure (kg)				
Seeds	900	300	400	1,600
Muna (provisions)	500	—	80	580
Fodder	—	800	800	1,600
Milling	180	60	60	300
Current expenses	300	—	—	300
Total	1,880	1,160	1,340	4,380
Annual crop (kg)				
Total produce	9,000	3,000	4,000	16,000
Less 'ushr (12.5%)	(1,125)	(375)	(500)	(2,000)
Less quarter due to labour	(1,700)	(250)	(300)	(2,250)
Less fixed expenditure	(1.880)	(1,160)	(1,340)	(4,380)
Net produce	**4,295**	**1,215**	**1,660**	**7,170**
Annual crop (piastres)				
Equivalent to	4,200	600	900	5,700
In Ottoman gold pounds				57

Table 8.3 Estimates of annual proceeds to Abujabers at al-Yaduda

	Per faddan		
	kg	Piastres	Gold pounds
Net produce from direct farming per each of 78 faddans	7,177	5,700	57
Net produce from farming on partnership basis being one-third crop of 18 faddans	2,700	2,200	22
Estimates of farming net income		*Totals*	
Total from direct farming	556,000	450,000	4,500
Total from partnerships	48,000	40,000	400
Total from sale of surplus animals, especially ewes, cattle and mares	—	100,000	1,000
	604,000	**590,000**	**5,900**

Prices of wheat were accounted for at six piastres per sa‘ or one piastre per kilogram, while those for barley, lentils, and other cereals were accounted for at half a piastre per kilogram, although they were generally higher, especially sesame and *hulba* (fenugreek).

The income of the Abujabers was high, considering the purchasing power of gold pounds, and was spent on hospitality in al-Yaduda's guest-house and on the Abujabers' three houses in al-Salt; on buying additional lands; building large houses; extended visits to Jerusalem, Nazareth, and Damascus; and on sending a few sons and daughters of the new generation to boarding school in Jerusalem. Farming in the frontier of settlement for Salih and his descendants has been and continues to be a rewarding operation.

9 Al-Nu'ayma: Ottoman Policy to Protect Tax-Paying Farmers

Immediately after the withdrawal of Egyptian forces from Transjordan in January 1841, the Ottoman government undertook to assert its influence in Jabal 'Ajlun, the northern district, which had until then been renowned for its unsafe conditions.[1] Not only were beduin tribes exacting the khawa from the villagers in the plains but also certain parts had thick forests that were the 'notorious home of bandits'.[2] It was in the government's interest to provide security for the peaceful inhabitants, who were normally the only payers of taxes.[3] The general feeling in governmental circles was that villagers were to be kept on the land by every means possible. Taxes, so badly needed by the poor wilaya of Damascus, could not be collected from deserted villages.[4] This policy was so fundamental, as far as the Ottomans were concerned, that it continued to be the cornerstone of their administration in Transjordan until their withdrawal in 1918. The use of force was therefore frequently resorted to in an attempt to achieve the aims of this policy.[5]

During this period, and even before the establishment of an Ottoman administration in the 'Ajlun area, the villagers of al-Wustiya petitioned the government concerning 'Arab al-Sa'aidi (al-Sa'aidi tribe) who had violently dispossessed them after many acts of aggression. Furthermore, the petitioners added, 'insecurity had reached such a level that they had decided to leave the country.' What followed is described by Gottlieb Schumacher, who in 1887 wrote:

> The Government however, at last, decided to attack the disturbers of public peace and sending out a sufficient number of cavalry with strict orders to exterminate the clans of highwaymen whose presence prevented the prosperous development of Ajlun, the soldiers for once did not neglect their duty. Beginning at Wadi al-'Arab, at this period the camping ground of 'Arab al-Sa'aidi who had violently dispossessed the proprietors of al-Wustiyih, they attacked them and with the aid of the villagers exterminated the whole beduin tribe down to its last member. 'The floods of Wadi al-'Arab were tinged with the blood of the Sa'aidi' as an eyewitness told me, to whom I owe the above account, 'and the corpses of the slain covered its slopes. We buried the enemy below those great isolated rocks of the Sa'aidi tribe.' In passing these rocks at a later date, I opened with a hoe the entrance to the caves below which were carefully built up with stones, and discovered within piles of human skulls and bones with pieces of rotten clothes and thus verified what my informant had told me. The villagers of nothern Ajlun are now left in peace, for the different clans of Bani Sakhr fled southwards, and the position of the beduins who remained has entirely changed.[6]

This severe act of discipline must have caused the whole population of northern Transjordan, fallahin and beduins alike, to heed the change in governmental behaviour. Within a few years, the taxes collected in the district grew to an extent that warranted the establishment of a governmental administration in 'Ajlun district in 1851. It was headed by a qa'immaqam whose seat was in the growing village of Irbid and whose province officially covered the whole area between the Yarmuk and Zarqa' rivers.[7] Gradually farmers became more numerous and the development of agriculture in the areas that were not yet under cultivation became more noticeable. A large group of farmers who came from Palestine, al-Gharayba (those from the west), settled in the five villages of Huwara, al-Mughayar, Kufr Jayiz, Irbid and al-Bariha.[8] Dayr al-Si'na had a population of about 100, mostly immigrants from Jabal Nablus, who settled there around 1875.[9] Another village was Makhraba, containing 45 well-built huts and about 250 inhabitants. They were immigrants from

Jabal Nablus and were all members of one family. They were very hard-working and, in addition to cereals both in the plateau and the Jordan Valley, they produced tobacco and excellent honey.[10] But farmers of the area and from Palestine were not the only settlers of the eastern frontiers. There were also the beduin tribesmen who changed their way of life and settled on the land, like the 'Anaza in the Jawlan, the Bani Sakhr and the smaller tribes who started to 'regularly settle down and themselves work the long despised dishonoured ploughshare, instead of investing the surrounding district with lances and levying a yearly tribute, the Khawa, as formerly'.[11]

During these days of continuous settlement, a different movement was taking place in al-Husn, the most eastern and largest village in Transjordan. In 1595 this village was noted in the Ottoman fiscal records as having a population of 24 households and 15 bachelors, all Muslims. The total taxes payable by the village, amounting to 17,153 akçes, gives a good idea of its high productivity and prosperity. Larger neighbouring villages paid less taxes; for example, Aydun (32 households and 21 bachelors) paid 10,215 akçes, and al-Sarih (58 households and 36 bachelors) paid 9,600 akçes.[12] At the time of Burckhardt's travels in 1812, al-Husn had grown to over 100 families, of which about 25 were Orthodox Christians under the jurisdiction of the patriarch of Jerusalem.[13] The Christian element in the village population was formed by families and clans that had come and settled in al-Husn mainly from al-Karak and the Hawran. They were well established in al-Husn and the leading house in hospitality in the village was that of a Christian. 'Abdallah al-Ghanim, of al-Ghanmat clan, was a prosperous farmer who was, in addition to his generosity, a pleasant host. The two travellers Seetzen (1806) and Burckhardt (1812) stayed with him in his house and gave the most complimentary reports about him.

The total area of the village is now 48,452 dunums,[14] to which should be added around 10,000 dunums in the neighbouring villages that were cultivated by the people of al-Husn. The distribution of these 300 faddans among the population is given in table 9.1.[15]

The question of leadership of the village during the 1860s was causing friction and instability; the two strong clans were

Table 9.1 The distribution of faddans among the population of al-Husn

Clan	Religion	No. of faddans	
Al-Numura	Christian	48	
Al-Nusayrat	Muslim		24
Al-'Amamira	Christian	24	
Al-Ayub	Christian	24	
Al-Hatamla	Muslim		24
Al-Maghayara	Muslim		18
Al-Rayahin	Christian	12	
Al-Ababsa	Christian	12	
Al-Ghanmat	Christian	12	
Different families	Mixed	20	82
Total (300 faddans)		152	148

contesting each other's right to be paramount. The oldest in al-Husn were al-Nusayrat, who claimed descent 15 generations back from a member of the large Ruwala tribe. They said that their forebear, Nusayr, left Tayma in northern Hijaz, with his two brothers, after a blood feud, and settled around 1600 in al-Husn. The two brothers, Nasir and Nassar, joined their tribe's encampments in the Hawran and their descendants, a Ruwala clan, are also called al-Nusayrat.[16] With an affiliation as strong as this one it is understandable that they claimed precedence in their area. After all, the Ruwala, their kin, were all over the Hawran and northern Transjordan during the second half of the nineteenth century and their influence was that of the largest beduin tribe in the whole of Syria.

On the other side there were al-Khasawna, who claimed descent from Muhammad Abu al-Fayd, a descendant of Abu Ja'far al-Sadiq and al-Husayn Ibn 'Ali, the grandson of the prophet Muhammad. Some time during the fifteenth century they migrated to southern Transjordan from the Hijaz, and after some movement in Palestine and Transjordan they settled in al-Karak. Towards the beginning of the seventeenth century, accompanied by a few other clans, they migrated to Jabal 'Ajlun after being defeated in tribal warfare with al-'Amr, who were then the shaykhs of al-Karak and its countryside. Within 50 years the area of Jabal 'Ajlun was taken over by Shaykh Dahir al-'Umar[17] who was paramount in Galilee and the neighbouring districts. Al-Khasawna found their opportunity for another fight and were in revolt against Dahir and his allies, al-Shrayda tribe. It is said that when Ahmad son of Dahir, who governed the area between 1771 and 1775, heard of their conspiracy with the Fuhayli tribe, he sent for their leader, Musa al-Muhammad, with the pretext of consultations. When he arrived, Ahmad ordered him to be beheaded and his head was hung at the gate of Tubna's citadel. He also ordered the confiscation of all Khasawna property. This development forced the tribe to migrate to Aydun, only 20 kilometres to the east but beyond the immediate control of Dahir and his son. When, a few years later, al-Jazzar, wali of Saida, defeated al-Zayadna, Dahir's tribe, al-Khasawna revenged themselves and plundered the property of their old enemies.[18]

Evidently al-Khasawna did not settle immediately in al-Husn, because by contemporary standards it was already heavily populated. Towards the 1850s, however, a few families settled in al-Husn and the friction started between them and the established shaykhs, al-Nusayrat. There followed what must have been some years of endless disputes and continuous unrest for the whole population. Al-Nusayrat needed their supporters among the other clans, especially the Christians, because al-Khasawna were more numerous and controlled a few villages in the area. The old enemies of al-Khasawna al-Shrayda of Jabal 'Ajlun also became involved in this running dispute and came to the aid of al-Nusayrat. They persuaded the Christian clans to send a delegation to Damascus and Istanbul to protest at the presence of al-Khasawna families in al-Husn.[19] A preliminary agreement was arrived at in 1285 H/1868 AD, during a session held by the council of Qada' 'Ajlun on 29 May 1868, but does not seem to have brought the case to rest. Therefore, the higher council of the liwa' al-Balqa' in Nablus was convened in 1286 H/1869 AD and a session was held specially to consider this important case. The text of the Nu'ayma document is as follows.[20]

> Notice has been taken of the exposition made by the Christians of the village of al-Husn in the Qada' of 'Ajlun which contends that they are being harmed by the residence in al-Husn of al-

Eviction order by the liwa' al-Balqa' in Nablus

Hindawi, 'Abd al-Rahman Hallush, 'Abd Allah al-Khalil, 'Abd al-'Aziz al-Nasir and his brother 'Abd Allah, who are committing encroachments against them. On gathering both parties and listening to their case in the Majlis, it appeared that if al-Hindawi and his mentioned relatives remain residents in the village of al-Husn, the disputes between the Christians and the aforementioned Muslims will not cease and that will be cause for the unrest of both parties and the destruction of the village.

After deliberation concerning that, and in the light of the memorandum of the Majlis of the Qada''Ajlun registered on the two-party undertaking dated 29 May 1285 [1868] containing the agreement of al-Hindawi and his party to the departure for the village of al-Nu'ayma on condition that three thousand and two hundred piastres (*ghurush*) required of them for the miri tax from the taxes of the district of Bani 'Ubayd will be added to the money (mal) levied on the village of al-Husn, thus freeing them of it, the recommendation is hereby made for the eviction of al-Hindawi and his aforementioned relatives from the village of al-Husn and their settlement at the village of al-Nu'ayma or the place they want in the Nahiya Bani 'Ubayd, on condition that they shall broadcast seeds in the fields they ploughed at al-Husn and drink the water in their wells for this year and that they will pay that which pertains to them of the miri taxes in the village whose fields they shall plough in equality with the population of that village. The population of al-Husn shall not be entitled to apportion on them any amount of the taxes of the village of al-Husn, nor of the three thousand and two hundred piastres mentioned before, after they have transported their produce from it. They will as of now transport their cereals and straw presently on the threshing grounds in al-Husn to the place in which they will reside. As to the cereals and hay which they have stored in the wells at the village of al-Husn, this they have undertaken, Muslims and Christians of al-Husn, to transport on their load-animals freely to the village that will be inhabited by al-Hindawi and his aforementioned relatives. Furthermore, when they broadcast their fields which are in the lands of the village of al-Husn and harvest their crops, they shall not stay overnight in al-Husn and neither shall they have an encampment in it, but they shall

> transport their produce to the village in which they shall reside and shall have their threshing grounds (baydar) in it. This memorandum has been writen in two copies with the agreement and consent of both parties, and should it be found agreeable to issue the exalted order for application as per the described arrangement, a copy should remain in the hands of each party. The command is with those who have the power to command.
>
> 29 Rabi' al-Akhir 1286

It is interesting to note that the chief factor during the session and later in the text of the decision was that of taxes. The main purpose was to remove the cause for the possible destruction of al-Husn, namely its being deserted by its own farmers. The document confirms that the deputy governor of the liwa' chaired the majlis, which was also attended by the Mufti (jurisconsult) Ahmad Abu al-Huda al-Khamash, the fiscal official, the adminstrative official and four council members. They all put their seals (not signatures) to the document, expressing the view that it was written with the satisfaction and acceptance of the two parties and subject to the approval of the higher authorities, usually meaning the governor in Nablus. Another interesting point in this document is that, although the Mufti endorsed it, the three religious leaders of the Orthodox Christians, the Latin (Catholic) Christians and the Samaritans did not see fit to attend the council. Evidently they were expected to be present because under their titles it is recorded that they were absent. Probably these three, by their absence, wished to convey to all concerned that the dispute was basically one between farmers concerning agricultural matters and which did not in any way involve religious differences. This becomes the more relevant when it is remembered that Nablus, like Damascus,[21] was not free from fanaticism and that it was only a decade earlier that there had been an uprising there against Europeans and Christians alike.[22]

The stringent conditions stipulating the arrangements for the final deportation as outlined in the document seem to have been what the Ottoman authorities wanted to apply, since the record was on the same day countersigned by the

Mutasarrif Muhammad Sa'id Pasha, governor of Liwa' al-Balqa', who added a notation at the top of the parchment.[23] His notation is short and firm: 'With its stipulations shall be made.' Judging by the results achieved, the order was very well implemented. Al-Hindawi and his relatives left al-Husn and for the following 118 years their descendants have lived and prospered in their estate at al-Nu'ayma. As far as it has been possible to ascertain, not one of the descendants of the deportees or any member of al-Khasawna clans lived in al-Husn after 1869.

The choice of a new home and farmland by Hindawi and his relatives was, by necessity, a limited one. They could only settle in the districts lying to the south and east of al-Husn. This was because the lands to the west and north were already settled by villagers and recent settlers from the western districts and Palestine. Besides, the deportees, probably more than 10 households, needed a few thousand dunums of dry farming land to make a living. Such an area was not to be found in the hilly country to the west in any case. Al-Nu'ayma met their requirements with its open fields totalling 34,484 dunums in area and its very small population. It had then only a few families of the Murayyan and Mawmaniya clans[24] and since there was a vast area of uncultivated fields, the newcomers were indeed welcome. The new settlers, it was hoped, would strengthen the hand of the fallahin in general against the beduins, especially of the Bani Sakhr tribe who were still claiming the khawa from the villages adjacent to the Badiya. They were actually correct in these expectations: within a few years, the fallahin of both al-Ramtha and al-Nu'ayma were defending themselves and their crops against their beduin neighbours in the east. The chronicles[25] relate who the new settlers were and how the lands were allotted to each household. The original partition must have taken place towards 1875: the fields of al-Nu'ayma were allocated in the first place; and a few years later, as more people settled down and their numbers grew, the fields of Kabar – an area of 11,204 dunums to the east of al-Nu'ayma – and Tumayra, an area of 25,577 dunums to the east of Kabar.[26] With 71,265 dunums in three estates there was enough agricultural land for all.

From the beginning, the total land of the village was divided into four parts. Each part was further divided into shares, each of which consisted of four faddans; such a share was called a *rub'a* in the north of Transjordan. Within each part the musha' system was applied, that is to say, the land was regarded as being the common property of all the farmers in that part, but the use of it was redistributed among them from time to time. This system afforded them the opportunity to co-operate more closely in the development of the fields through better ploughing, removal of stones and applying more balanced crop rotations. Had the musha' system involved all the farmers in the village, it would not have been as easy for them to co-operate as the rotation would have become much more diverse. When, after a few years, farming extended to Kabar and Tumayra, the numbers had grown to such an extent that it was not possible to apply the system of al-Nu'ayma and therefore all the fields in these two villages were run under the musha' system. This meant less work and less income for the farmers and, for the fields, less development and a lower standard of agricultural care.

The four groups were the following:

1 Rub' al-Shuyukh or Rub' al-Khasawna. Although Hindawi was the leader of this group in al-Husn, al-Khasawna were the recognized leaders of the clan. A number of them from Aydun and the neighbouring villages joined 'Abd al-'Aziz and his brother 'Abd Allah. Since they were recognized as the shaykhs, the title was surrendered to them without contest and is used until this day. This rub' belonged to al-Khasawna alone.
2 Rub' al-Hindawi, who ran 24 faddans with al-Murayyan clan who were in al-Nu'ayma prior to 1868. As his partners, Hindawi had the clans of Barakat, Naqarsha, Shawamla, Abu Zayd and 'Asi.
3 Rub' al-Sumadiya, who also had as partners a few small clans. This tribe is considered one of the most numerous in Jabal 'Ajlun. They claim descent from al-Husayn Ibn 'Ali and are spread throughout Syria, Palestine and Transjordan. Probably they joined al-Khasawna when they moved eastwards into al-Nu'ayma.

4 Rub' Hallush, whose partners were the Hamashat, Mawmaniya and Shutnawiya. 'Abd al-Rahman Hallush was one of the original group deported from al-Husn. The Mawmaniya, a large tribe and related to al-Sumadiya, were in al-Nu'ayma prior to 1868, while al-Shutnawiya originally joined al-Khasawna in their migration from al-Karak in the seventeenth century.

The new settlement must have needed a few years before the agricultural system became truly established, but from the outset Hindawi was an outstanding leader. Having lost the reasons, and perhaps the appetite, for fights and disputes, he seems to have settled down and become an especially good farmer. He had the largest agricultural operation in the north of Transjordan and southern Hawran. Like his colleague, Salih Abujaber at al-Yaduda, he imported his harathin from villages in Jabal Nablus and Hawran. He also gave them one-quarter of the crop and maintained them in the household during the agricultural year. He was producing in good years somewhere between 200 and 300 tons of cereals, the surplus of which he sold to the beduin tribes who visited the area in summer, or exported to Acre in Palestine. Just before the turn of the century, he built his big house to which was attached a large courtyard and bayika which is one of the largest in the northern district, if not the largest.

The settlement in al-Nu'ayma was, of necessity, an extension of the confrontation that had existed for centuries between the nomads and the farmers. Supported by the Ottoman government, which deported him and his relatives from al-Husn in the first place, Hindawi, like Sattam Ibn Fayiz, must have also felt the winds of change. He therefore co-operated with the government whose main aim was the collection of a greater yearly amount of taxes. Al-Nu'ayma started paying the tithe to the qa'immaqam of 'Ajlun immediately after the one-year exemption from tax payment awarded them during the last season in al-Husn. In return for this, the inhabitants of al-Nu'ayma received the uncontested right to the land as well as protection, in different degrees, against beduin incursions into the area.

Being the furthest eastern village in the frontier of

settlement, Al Nu'ayma had to bear in mind two important considerations. The first was that, to maintain the lands taken over, they had to increase their numbers. This was necessary in order to impress on their neighbours and visiting tribes their ability to fight any incursions on their rights, and to have the manpower for the cultivation of the fields that the Ottoman government seemed so interested in accomplishing. To achieve this aim, Hindawi welcomed newcomers from all directions but especially from his clan, the Khasawna. He encouraged them to come and settle, and in this regard we have one case on record when a clan from al-Karak, the sons of Jiryis al-Halasa, left their town in 1870 and moved northwards to al-Nu'ayma. When they arrived, they were assisted by Hindawi, the shaykh of the village, in acquiring what agricultural land they needed and building a house. They returned to al-Karak in 1925[27] apparently because of bad economic conditions encountered by farmers in general at that time and a clannish yearning to be reunited with their tribe in al-Karak.[28] The policy of encouragement adopted by Hindawi towards newcomers brought in good results, and the growth of population in al-Nu'ayma was outstanding: by 1979, it had 791 families and a total population of 5,950 people, divided into 3,102 males and 2,848 females.[29] The growth basically happened in the Transjordanian sector of the population, al Nu'ayma receiving only a minimal increase from Palestinians after the two migrations of 1948 and 1967.[30]

The second important consideration was that, in spite of the government support and protection in moments of need, the beduin threat was a serious fact of life that had to be tackled. Hindawi followed two courses to achieve his aim in these difficult circumstances. He concluded alliances with his neighbours in the north and the west to provide joint defence whenever possible against the marauding tribes from the east. Chief among these allies was the population of al-Ramtha who, in addition to their fields, owned the fields of Umm al-'Abar and rebuilt their village after it was ransacked and completely destroyed by the joint attack of the Bani Sakhr and the 'Adwan in 1869.[31] His flank in the south was relatively well protected since it was the start of the hilly country that did not provide the beduins with any special pastures for their herds and flocks. The second course involved, in contrast with the show of strength, a show of continuous hospitality. He had an open house in his quarters in the village centre and every group of visitors was pleasantly entertained and offered all the requirements of the famous Arab hospitality. This attitude gained for al-Nu'ayma and Hindawi the goodwill and easy relations that they badly needed, especially among the beduin nomadic tribes to the east, and gave the village a reputation for hospitality and generosity. In Transjordanian society of the 1890s this was indeed a worthy investment that paid handsome dividends.

The fertility of the new fields in these new settlements succeeded in turning Hindawi, his relatives and the newcomers into hard-working fallahin, owning the land and producing excellent crops. The success of the venture also opened the eyes of some of the beduin shaykhs to the possibilities inherent in land ownership and cultivation in general. Al-Khurshan shaykhs of the northern section of the Bani Sakhr laid claim to Khanasiri and Fa',[32] both to the east of Tumayra, and the tribe owns them to this day. Another tribe, Bani Khalid, whose chief village is Hawshah, took over the lands north of Khanasiri and Tumayra which they also own to this day.[33] Some 25 years after him, the beduins of northern Transjordan were following in the footsteps of that gentleman-farmer Sattam. Gradually, they too realized that with a stronger government willing to resort to force they were better off as landowners.

Before his death in the year 1320 H/1902 AD,[34] Hindawi had seen his dream come true. He had four sons: Wanas, Muhammad, 'Ali, and Salim. 'Ali was killed during fights with the beduins and left no children. The remaining three brothers established with their descendants the Hindawi clan that became one of the most hospitable and prominent in northern Transjordan. Both as farmers and as leaders – some of them having been deputies, ministers and ambassadors – from after the first world war until the present, they have served their country well.

The role that al-Nu'ayma played in the extension of the frontier of agriculture since the 1870s cannot therefore be underestimated. Together with its sister villages on the edge of the Badiya, it has maintained a tradition of perseverance and courage that has brought for its people not only material reward but also the satisfaction – which other communities felt before them – of the knowledge that being a farmer does not necessarily mean being a second-class citizen, as beduins generally believed.

10 Sattam Ibn Fayiz: a Gentleman Farmer among Nomads

The first reports about the involvement of a purely nomadic tribe with agriculture in Transjordan concern either Sattam[1] or his father, Fandi Ibn Fayiz,[2] towards the end of the 1860s. These reports differ as to who was the pioneer. Peake, in 1928, mentioned that Fandi, shaykh al-mashayikh of the Bani Sakhr tribe, started the cultivation of cereals on a small scale in the fields of Khirba Barazayn to the south of al-Yaduda. In another place Peake said: 'Sattam Ibn Fayiz was the first in Bani Sakhr who gave attention to agriculture and cared for it. He possessed rich lands in al-Balqa.'[3]

Both Shaykh 'Awad al-Sattam[4] and Shaykh Ghalib al-Nawwaf[5] confirmed during interviews with them on 1 and 8 August 1985 that it was Sattam who started cultivation but that Fandi, seeing the plentiful crop in Barazayn, could not resist taking it for himself. This site was then within the area of the Balqa' confederation and the cultivated fields could have been acquired by the Bani Sakhr shaykhs as dowry, following the prevailing custom.[6] At this time, the two confederations were at peace, as a result of intermarriage and also the feeling that both the 'Adwan and Bani Sakhr were being carefully watched by the Ottoman authorities. Sattam was taken and imprisoned in Nablus,[7] and a few years before that, between 1863 and 1865, Hulu Pasha, who assumed the governorship of the sanjaq of Nablus, took 'Ali al-Dhiyab[8] and imprisoned him in Acre.[9] It is certain that agriculture was practised in the lands of both confederations in an attempt to please the government and also to compensate for the loss of income that until then they had derived from khawa and *ghazu* (raiding) and, in the case of Bani Sakhr, from the pilgrimage caravan.

Sattam, who had eleven brothers and eight sons, was capable, progressive and definitely ahead of his time. He had on many occasions visited Damascus, Jerusalem and Nablus; and before 1870, according to Tristram, he visited Alexandria and Cairo, when he took some Arab horses as a present to the Khedive of Egypt. During that trip he journeyed in a railway train and saw for himself how people lived in a settled society like Egypt. He married at least five times and in every instance his prestige was increased by his in-laws, who were all influential shaykhs:[10]

		Sons
1	From al-Zabn, the strong clan of Bani Sakhr who for a long time declined to acknowledge the chieftainship of Ibn Fayiz	Fayiz Fawwaz
2	From al-Ruwala, the celebrated tribe in Badiya al-Sham	Mithqal
3	From al-'Adwan, 'Alya sister of Shaykh 'Ali al-Dhiyab	Nawwaf
4	From al-Sardiya, shaykh of the nomadic confederation of northern Transjordan	Jurayid Shibli 'Arif
5	From al-Khurshan, the second faction of the Bani Sakhr	'Awad

The genealogy in figure 10.1 attempts to convey the relationship between the names that appear in this chapter. Shaykh 'Awad and Shaykh Nawwaf, as well as other dignitaries, have kindly helped in recording the different facts and dates.[11]

In addition to Sattam's love of and desire for agriculture he also faced a challenge that made him become more involved with it. This concerned the difference already mentioned between him and his father Fandi. According to Mithqal Pasha al-Fayiz, son of Sattam, and shaykh mashayikh Bani Sakhr between 1921 and 1967, Sattam was indeed the pioneer. He evidently cultivated the field in Barazayn without the

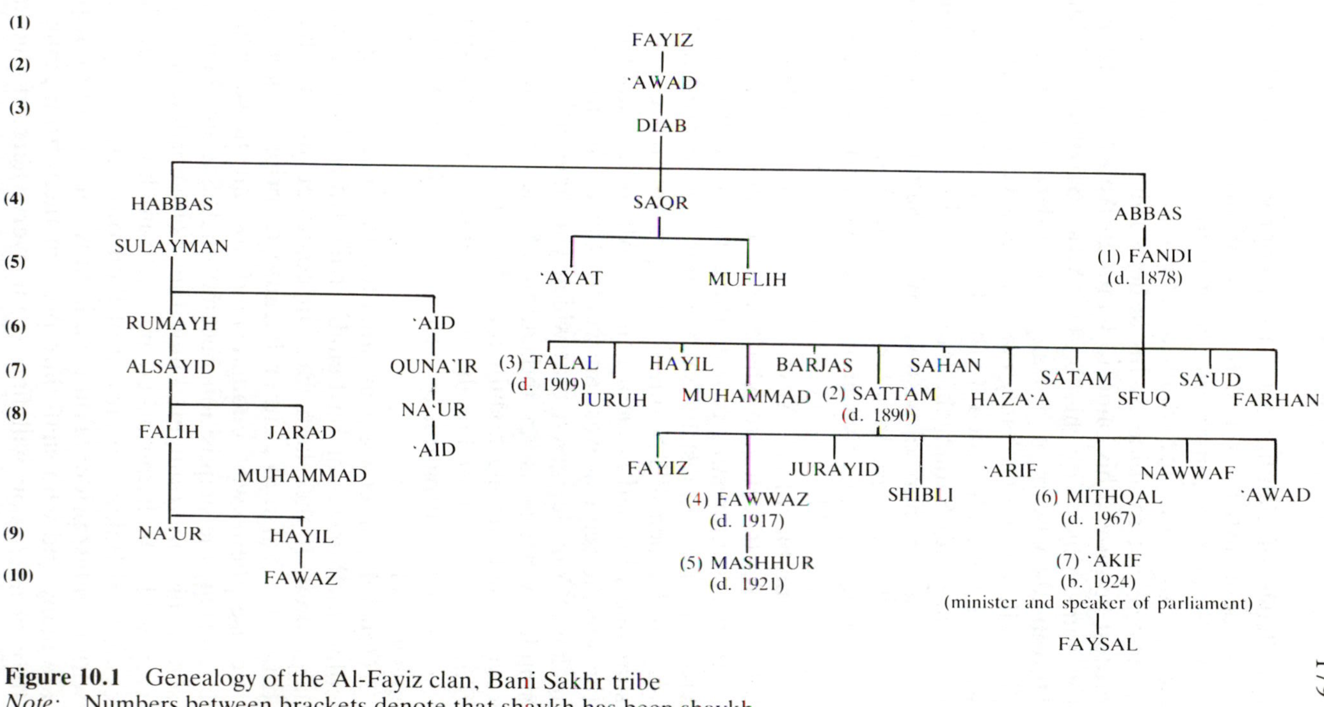

Figure 10.1 Genealogy of the Al-Fayiz clan, Bani Sakhr tribe
Note: Numbers between brackets denote that shaykh has been shaykh al-mashayikh of Bani Sakhr. Numbers 2, 3 and 4 have been governors of Jiza (Zizia) during the Ottoman period.

knowledge of Fandi,[12] who then appropriated the crop for subsequent distribution to the tents of his sons, cousins, and clans. Sattam remonstrated but was told that his father had eleven other sons for whom he also had to care. Sattam had no choice but to abide by his father's ruling, but as a sign of protest he moved his camp to Umm al-'Amad, where he lived for some time with his bride 'Alya, daughter of Shaykh Dhiyab al-'Adwan of al-Balqa' confederation. She was his third wife and he probably married her a year earlier, in 1872, when she was 'reputed to be the belle of the country'.[13] Fortunately for Sattam she had, in addition to her outstanding beauty, a fair share of wisdom. When Sattam moved his tent and small encampment to Umm al-'Amad, she counselled him to visit her brother 'Ali and ask for land to cultivate. When approached, 'Ali was generous, as would be expected of shaykh mashayikh al-'Adwan to his brother-in-law, and he told Sattam to cultivate whatever lands he could.

It is very interesting to note the expressions used during the visit. For the request by Sattam, the Arabic word *al-hadhiya* was used which is a 'request for a present that is asked from someone who had made a gain'.[14] 'Ali, before knowing what the specific request was, answered 'Ibshir billi jit bih', which means, 'Consider given that which you have asked for'. This simple exchange is an indication of the prevalent attitude current then among beduin regarding the ownership of land. Since it was neither bought nor inherited, it was acquired by assumption, usurpation or occupation by force, and thus was given freely.

Generally a claim was announced by a strong shaykh, either for himself or on behalf of his tribe, to a khirba or a large area. If the claim was not contested by any shaykh or tribe of equal rank, either because of prior arrangement or because there was enough land for all those who were interested, the acquisition became final. Sattam, being a true son of his environment, knew these circumstances and rules very well. On being asked by Tristram how his people had come to al-Balqa', so far out of their own territory, he replied with manifest pride: 'True, the land is not ours but our people are many and who shall dare prevent them from going where they please? You will find them everywhere, if the land is good for them.' Tristram commented that this was one of the many advantages of belonging to the strongest, where might is right, and Sattam was the shaykh of the strongest tribe in Transjordan.[15]

Sattam did not fail to seize the great opportunity that was presented to him by his brother-in-law. Accompanied by only two companions he rode to the mound at Umm al-'Amad and fired a shot into the air, confirming that it had become his and as of that moment he was its protector. Without delay, he rode in succession to al-Zabayir, Zizia, Zuwayzia, Umm Rummana, Manja, Jalul, Huwara, Umm Qusayr and at the end of a long day's ride arrived at Dulayla.[16] In every one of these khirab, he erected a mound of stones, fired a shot into the air and declared the site his own property. Thus all these ten villages came into his private ownership, while his eleven brothers remained with their father at Barazayn. Later he found it more expedient to distribute some of these lands to his brothers and those from among the important clans of the Bani Sakhr with whom he wished to ingratiate himself. He gave al-Zabayir to Sahan; Umm Rummana to Muhammad, Hayil and Juruh; Manja to the Kunay'an clan; Jalul to the Zabn clan;[17] Huwara to the Shulash clan; Umm Qusayr to al-Nufal clan. Later, in 1881, he gave Dulayla to 'Ayd al-Rudayni in compensation for Madaba which the Ottoman government had by then allotted to the three Christian tribes who migrated from al-Karak (see chapter 12). In addition to Barazayn, he kept Umm al-'Amad, Zizia and Zuwayzia for himself, his younger brothers and growing sons. A prosperous village now, al-Qastal, which belongs to the descendants of Sattam, must have been considered as Bani Sakhr domain, but was not cultivated until after the turn of the century. Gertrude Bell, passing through the area in May 1900, wrote that al-Qastal was uninhabited and the land around it uncultivated.[18]

To be able to appreciate the importance of this development, it is necessary to acquaint oneself with the vast area that was involved as well as the economic potential of these many fields of which ownership was acquired in a relatively short time.[19] The area of each of the ten khirbas, in accordance with the information available from the records of the Department of

Table 10.1 The areas of the 10 khirbas

Khirba (village)	Remarks	Area in Dunums
(1) Umm al-'Amad	The seat of al-Fayiz shaykhs. The name means 'Mother of columns', a reference to its ruins	16,379
(2) Al-Zabayir	At present there are two: Zabayir al-'Adwan, the name reflecting its ownership of a century ago; and Zabayir al-Tuwal, referring to a clan of Madaba who bought lands in it	12,328
(3) Zizia	Also called al-Jiza, probably by the Egyptians during their occupation of Syria 1831–41. An important Hajj and railway station, it was the seat of the nahiya founded in 1881 under Shaykh Sattam	39,614
(4) Zuwayzia	The word is a diminutive of Zizia	17,094
(5) Umm Rummana	At present owned by descendants of four brothers of Sattam known as al-Dhiyab. The name means 'Mother of the pomegranate'	15,875
(6) Manja	The name means 'Refuge'. Property of the Kunay'an clan	9,104
(7) Jalul	Still the seat of al-Zabn clan	16,800
(8) Huwara	The name means 'Poplar tree'. Property of the Shlash clan	9,912
(9) Umm Qusayr	The name translates as 'Mother of a small palace', meaning the ruin of an estate house. Property of al-Nufal clan	17,231
(10) Dulayla	Known as Dulayla al-Mutayrat to distinguish it from its neighbour in the south belonging to the tribe of Banih Hamida and known as Dulayla al-Hamayda	15,523
Total area		169,860

Sources: see text.

Lands and Surveys of the Jordan government, is outlined in table 10.1. Further notes that are relevant have been added. The total area of land, 169,860 dunums or 17,000 hectares, in addition to being strategically located – since it controlled the Hajj route from Damascus to the holy places and formed the buffer between the cultivated districts in the west and the fringe of the desert – offered excellent scope for agriculture and pasturage. The fields in these khirbas, except in the eastern part of Zizia, are large, slightly hilly and their thick red soil is best suited for the production of different types of cereals, especially barley and hard wheat. The annual yield, unless there was a drastic shortage of rainfall, was good, in general varying between 10 and 20 sa‘s to the measure. Such a crop, in the 1880s, must have considerably improved the quality of life for the southern clans of the Bani Sakhr who were already enjoying a higher than average standard of living, with their 12–15,000 camels.[20] The development of agriculture led to an improvement for all nomadic tribes dependent on a pastoral economy, though tribes like the Bani Sakhr and the Ruwala, who may have had nearer 80,000 camels, were already much richer than say the Shararat or Hutaym of wadi al-Sirhan and northern Hijaz who were considered poor.

This strong desire on the part of a nomadic shaykh to own large estates and yet have the ready willingness to distribute many of them, with such largess, clearly reveals an outstanding personality. Nothing could exceed the dignity and stateliness of this young seigneur[21] whose manners were above criticism, and who treated chieftains and dependants with dignified superiority.[22] His attitude, being the result of wisdom and far-sightedness, signified the vanguard of a new trend in beduin social and economic behaviour, throughout al-Balqa'. This was the readiness among some nomads to settle, after it became clear that agriculture could be a new source of wealth. The tendency must have been enhanced, over the following few years, by the realization that the central government, at long last, was serious in its efforts to control the countryside.[23] During the 1870s, beduin shaykhs as well as headmen of settled communities gradually became aware of this important development and their apprehension of its effects was increased by the drastic change brought about in the equipment of the government forces. The armament of some Ottoman units, starting in 1868, improved greatly when some battalions were issued with Snider mechanisms.[24] This was a system adopted in England during 1867 for converting the Enfield rifle following a design by Jacob Snider of New York. The conversion allowed the muzzle-loader rifles to become breech-loaders, capable of using completely self-contained metallic cartridges which the user could carry with him in any quantity he chose.[25] This new situation, when beduins had only muzzle-loaders, enabled the government forces to muster a new level of fire-power and helped to bring about greater respect for the *dawla* (government) and its armed forces.

Sattam seems to have been aware, more than the shaykhs of other tribes, of the meaning of the new presence in al-Salt, some 30 kilometres to the north-west of his estates. The qa'immaqam who had been instated there a few years before had at his disposal a body of regular and better-equipped troops, not to mention the irregulars who would rally to his call.[26] He could also count on reinforcements from the sanjaq of Nablus to which he was attached. Furthermore, it was evident that, since the spread of governmental authority, the settled population were in support of the new drive for stability and security. They started to pay taxes regularly to the government and the khawa system of payment to beduin shaykhs was more frequently contested. In 1881, even Sattam himself was rebuffed when the settler-farmers in Madaba flatly and rather crudely refused to pay him any khawa, saying that they were neither ready nor willing to pay any amount over and above that which they had to pay to the government in al-Salt. Wisely, he did not try to force the issue.[27]

Another factor that strongly influenced the Bani Sakhr was their relationship with their two rivals, the 'Adwan and the government itself. 'Ali al-Dhiyab, and before him his father Dhiyab, had been imprisoned by the governor of Nablus and yet the 'Adwan in recent times had been willing to assist the government, to the apparent detriment of the Bani Sakhr. Fandi, Sattam's own father, had great difficulties in regaining the favour of the government in Damascus. After three years of great loss in prestige and money, the Bani Sakhr were reinstated as supervisors and camel-suppliers to the pilgrimage caravan, and Fandi set out on 27 February 1872 with a strong detachment consisting of 700 of his camel men to accompany the caravan from al-Ramtha in the north to a distance six days ride south of al-Karak.[28] In addition to the usual *sarr* (purses) that were presented by the pasha of the pilgrimage to Fandi and his accompanying shaykhs, the caravan normally hired hundreds of camels at a few gold pounds each. Other benefits were realized through the rendering of different services, including the sale of provisions, at the stations of Zizia and Qatrani. The economic importance to all al-Fayiz and the whole Bani Sakhr accruing from good relations with the government was therefore very much appreciated by Sattam, and a few years later became the corner-stone of his policy.

Governmental authority continued to develop and the friendly trend among the nomadic tribes similarly continued to improve. In this connection one is reminded of the great truth contained in the writing of Carleton S. Coon, who said:

> The political situation of the beduins varies from period to period. When the central governments to which the tribal territories are officially assigned are weak, the paramount shaykhs rule virtually as Kings[29] and even cities have paid them tribute. At times, when the central governments are strong, their authority becomes purely local.[30]

Sattam's prestige and good public relations grew over the years and it is not surprising therefore to hear of him as the mudir (governor) of the newly created Nahiya al-Jiza (Zizia) in 1881. The Salname Vilayet-i Suriya for the year 1299 H/1881 AD mentions Sattam as the mudir of 19 villages at a time when the whole qada' of al-Salt, including the only other administrative unit, the nahiya of al-Jiza, had only 25 villages in all.[31] Other than the large number of villages under his care Sattam had also, as of 1881, the task of being shaykh al-mashayikh for the whole Bani Sakhr tribe, an estimated 15,000 people in all (3,000 tents).[32] Major C. R. Conder was making his famous survey in al-Balqa' during October 1881. He was not well assisted by the Ottomans in the area and when his presence was revealed wrote in a spirit of helpless anger: 'Sattam was on good terms with the governor of al-Salt and he gave information to Muhammad Said Pasha, while conducting the Haj, of the presence of the survey party in the 'Adwan Country.'[33] Thus alerted, the governor of al-Salt acted to prevent the survey. Sattam's duty was to maintain order in the whole area and, ex-officio as the Bani Sakhr paramount chief, he had to assist the government in the safe conduct and return of the Hajj caravan through the area between al-Ramtha and Qatrani, a distance of nearly 225 kilometres. The governorship of al-Jiza continued in the house of al-Fayiz. The Salname of 1891 stated that Talal Effendi al-Fayiz was acting governor, while the two Salnames of 1895 and 1896 stated that he was the governor. Talal, a brother of Sattam, ruled until 1909[34] and was followed in the *mashyakha* (chieftainship) as well as the governorship by Fawwaz, son of Sattam, then by Mashhur, son of Fawwaz, who was killed in 1921[35] (see figure 10.1, p. 179).

Unlike al-Yaduda, at Umm al-'Amad, Sattam's main village, there was evidently not the ability or know-how to develop a system for large-scale agriculture. It therefore had a small number of faddans that were run directly on Sattam's own account and the major activity was left to the farmer-partners who, having nothing better to do, rallied to the new opportunity. They were mainly Egyptians, some probably descendants of soldiers who deserted from Ibrahim Pasha's army during his evacuation from Syria in January 1841.[36] A good part of the work, however, must have been performed by the slaves owned by the shaykhs of the Bani Sakhr. 'Ayd al-Rudayni, who was compensated with Dulayla by Sattam after Madaba was cultivated by the Christian tribesmen in 1881, took care in 1872 to point out to Tristram and his group that no Bani Sakhr tribesman ever drove a plough himself, but left all such menial work to his slaves.[37] This may have been true until 1880 but it would have been against all economic principles if it continued after that, when raiding, plunder, and the khawa became increasingly difficult to perform under a more disciplined and exacting government. This does not mean that instability and raids did not continue in the eastern areas, but rather reflects the fact that Sattam, as governor of al-Jiza, was performing his duties in a most satisfactory manner; order really did prevail in the villages between al-Salt and Zizia after he took over his duties. This was indirectly confirmed by Shaykh ‘Awad al-Sattam, who said that even until 1918 when the first world war ended, any Sukhari (one of the Bani Sakhr) who traversed the railway line to the east of the villages of the Bani Sakhr was out of the jurisdiction of the dawla, meaning that no troops would be sent after him into the desert but that the government was in full control to the west of that line.

The 30-odd faddans that Sattam ran on his own account at Umm al-‘Amad must have been developed gradually. At the start of his farming activity he did not have oxen or mules, and therefore had no choice but to allow the use of some of his camels as animals of toil. Oxen for ploughing were not available as cattle were not kept by the tribes; mules were not raised and horses were too dear by beduin standards to be employed in what was then considered menial labour. However, through his contacts with the beduin tribes of eastern Palestine around Gaza and Bi’r al-Sabi‘ (Beersheba), Sattam must have noticed that camels were trained to work in agriculture.[38] This continued for a few years until one day after his return to Umm al-‘Amad from Damascus he became very angry when he noticed the condition of his camels’ feet.[39] He raged at his farm manager, an Egyptian by the name of ‘Abd Allah Abu Zayd, who was able, after some difficult moments, to explain that the camels’ swollen feet were the result of their being employed in threshing the abundant quantities of wheat in a year when there had been a very good harvest. He added that this would not have happened if they had mules on the estate.[40] Sattam then became determined to effect a change and swore to provide his murabi‘iya with both mules and oxen, which he later bought from the villages of Irbid and the Jawlan. He further allowed the use of ordinary horses after they had been castrated (known then as *kudsh*, singular *kadish*) in agricultural tasks.[41] Once more the shaykh al-mashayikh, outstanding and practical as he was, saw no harm in changing course and adapting himself to a new situation.

Besides his 30 faddans, Sattam must have had many more on a partnership basis with the fallahin in his different villages. In addition to the Egyptians, there were also among them members of poorer tribes attached to the Bani Sakhr, farmers from al-Salt, the neighbouring tribes, Hawranis, and Palestinians. The Egyptians, who had formed into clans by 1890, were in three groups:

1. Al-Zuyud (sing. Abi Zayd) were headed by Shaykh 'Abd Allah Abu Zayd, who as a highly respectable man was also acting as farm manager and agricultural counsellor to Shaykh Sattam. They had around four faddans between them.
2. Al-Maharma were headed by Husayn Hasan and they also had around four faddans between them.
3. Al-Taharwa were headed by Shaykh Muhammad al-Taharwa who was a religious man. They had around five faddans between them.

The other fallahin partners were two or three families from Hawran, who were all called Hawranis even though they may not even have been from the same village. They ran about three faddans between them. Al-Qaysi from Samu'u village in the Hebron district was running about ten faddans. Two or three families from al-Salt were also running three faddans in Umm al-'Amad.[42] The total area under cultivation could have therefore been around 60 faddans in all.[43] That meant about 12,000 dunums at the rate of 200 dunums to the faddan. The fallow system was predominant and therefore only half this area would be under cultivation in any one year.[44] Until the death of Sattam cultivation was limited to the most westerly areas of his estate in Umm al-'Amad, as they were the best and received the highest rainfall. His lands in Zizia and Zuwayzia were only cultivated by his sons after the turn of the century. The water supply for Umm al-'Amad, as in the whole plateau of al-Balqa', was a chronic problem. The Bani Sakhr had of course the advantage of having the big pools in Zizia and al-Qastal, which were maintained and repaired by the Ottoman authorities for use during the passage and return of the Hajj caravan, but the water was seven kilometres away in al-Qastal and double the distance in Zizia and was therefore only resorted to in case of dire need.[45] In ordinary times the whole household in Umm al-'Amad drank from the old wells that were cleaned by Sattam's men and fallahin and were used for water collection. It has not been possible to ascertain whether new wells were dug by Sattam (as they were at al-Yaduda), but this is unlikely since even in the 1930s Umm al-'Amad had to bring water on camel-back from the running water of Hisban, ten kilometres to the west, during drought years.

The crops produced on Sattam's estate were, in order of importance, wheat, barley, millet, lentils and chick-peas. Wheat was by far the major crop and perhaps accounted for two-thirds of the cultivated area. Barley was also important because the Bani Sakhr's prestige depended on their great number of mares which may have exceeded 150 in the encampments of al-Fayiz alone. Millet was planted because of its high productivity and was used instead of wheat for bread by the poorer class or occasionally by everybody in drought years. Lentils were used by the Bani Sakhr to make soup but they did not care for chick-peas which, as 'Awad al-Sattam confirmed, were planted by the fallahin and generally sold to merchants in towns. The net produce every year, amounting to between 200 and 300 tons, was used by Sattam in his own encampment and those of his nearest kin. There must have been a surplus since these encampments could not have numbered, together with the fallahin in the different villages, more than 2,000 people, who would have needed around 200 tons of cereals at the most, although more would have been consumed by Sattam's many guests.[46] Probably the surplus was sold to the other clans of the Bani Sakhr as well as to friendly tribes visiting the area. It is possible that some of the surplus wheat found its way, through farmers and merchants, to al-Salt and thence to Jerusalem and Nablus, but it is certain that neither Sattam nor his sons ever handled an export operation to the western districts during the nineteenth century.

Although the importance of Sattam's agricultural venture rests mainly on the fact that he was the first nomadic shaykh not only to own land but also to place it under cultivation, it also has an economic side. Through the production of their own requirements of cereals, Sattam and his encampments avoided the drain on their other resources; there was no longer any need to purchase provisions. Having become producers themselves they not only saved what they would have spent but started as well to earn extra income from the sale of the surplus products. The following statement, although based on very rough estimates, will provide a clearer picture of the economics of the venture as a whole.

There were 30 faddans run on Sattam's own account that needed about 6,000 sa's of wheat and other cereals for seeds; this meant a total seed quantity of nearly 36,000 kilograms.

If a medium-crop year is taken as average, then these seeds would give 540,000 kilograms as yield. Of this the murabi'iya received one-quarter for their work during the agricultural year and this left for the estate 405,000 kilograms. Another 36,000 kilograms were needed for *bidhar* (seeds) and 40,000 kilograms for muna. The 'ushr amounted to around 50,000 kilograms. Taking into account general expenses this left the estate with 269,000 kilograms:

	kg
Produce at the rate of 15 to the measure	540,000
Less:	
Share of Murabi'iya	(135,000)
Bidhar (seeds)	(36,000)
General expenses	(20,000)
Muna (provisions)	(30,000)
'Ushr (tithe)	(50,000)
Net crop in one year	269,000
To this must be added the estate's share from the gross crop of the other 30 faddans run by the fallahin partners at the rate of one-third of the crop for the landlords	180,000
Total net crop in one year	449,000

At the ruling prices for cereals in the 1880s a total gross crop of 450,000 kilograms could realize between 225,000 and 250,000 piastres when the rate was 120 piastres to the gold

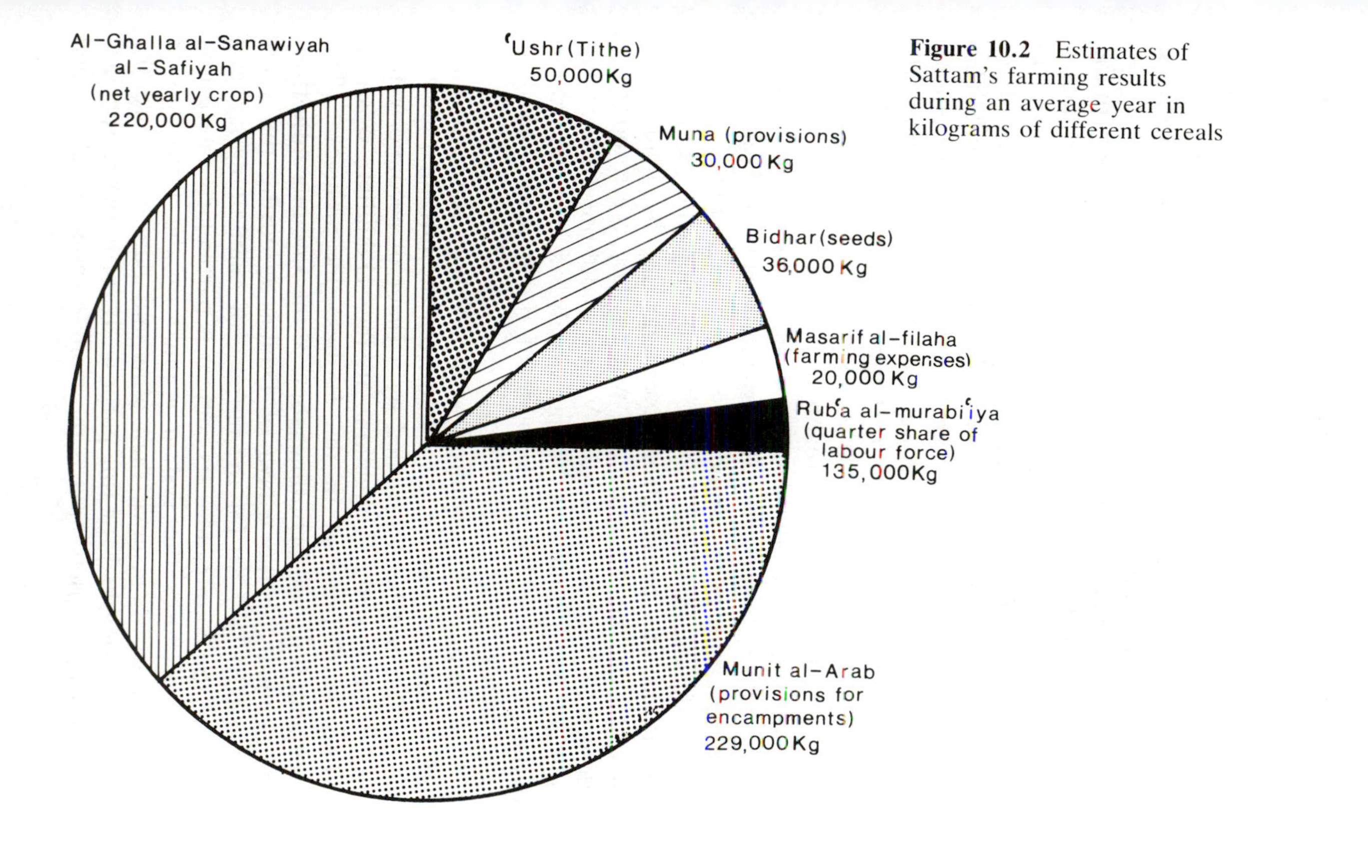

Figure 10.2 Estimates of Sattam's farming results during an average year in kilograms of different cereals

pound sterling. The requirements of the 2,000 persons in Sattam's encampments as well as those of his many guests probably amounted to more than half that quantity. The rest, some 200,000 kilograms of wheat valued at nearly £1,000 in gold, represented an impressive sum in those days. Figure 10.2 shows that farming, with very few out-of-pocket expenses, was indeed a profitable operation.

Sattam's pioneering work in the field of agriculture with the formerly nomadic Bani Sakhr greatly improved the standard of living among their clans. It not only did so through the production of different cereals but had a significant side effect which seems to go hand in hand with settlement on land. This was the important development of animal husbandry among the Bani Sakhr tribesmen as well as their farmer-partners. By the time of Sattam's death in Umm al-'Amad in 1891, his encampments had over 5,000 sheep and goats and nearly 500 cows. In addition to this wealth he and his nearest kin had probably 2–3,000 of the camels owned by the whole tribe of Bani Sakhr. These herds, the original wealth and source of boundless pride of the clans, were only brought to the western ranges during May and June so as to allow the camels the opportunity to benefit from pasturage in the fields after harvest. During autumn and winter they were normally allowed to graze in the Jordan Valley where the weather was mild and water abundant. Spring was always their time in the Syrian desert, the original abode of the nomads and their camel herds.

In addition to his acquired wealth, Sattam seems to have had a substantial invisible income from the Ottoman government. This has been revealed by a sheet of accounts which was found in the collection of the Dakhiliat Wazarati (Ministry of Interior) in the Başbakanlik archives.[47] The document (table 10.2) outlines some of the amounts authorized for payment by the wilaya of Damascus during the years 1291–4 H/1874–7 AD for the purpose of maintaining law and order in the countryside. Its content is particularly interesting since it discloses the system applied by the Ottomans in certain areas to achieve their aims and puts on record the amounts earmarked for this purpose. It furthermore specifies the outlying districts, that is the trouble spots, which needed

Table 10.2 Document of Ottoman government accounts for the Sanjaqs of Damascus, Hawran and Balqa', 1874–7

District	Ranks	Monthly salary	Number of persons	Total piastres monthly	Total piastres yearly
Sanjaq of Damascus					
Hasbaya, Majdal Shams and environs (Jawlan and southern Lebanon)	Yuzbashi (captain)	400	1	400	4,800
	Soldiers	160	30	4,800	57,600
	Total	—	31	5,200	62,400
Dayr 'Atiya and Dumayr (Ruwala domain north-east of Damascus)	—	150	25	3,750	45,000
Jabal Qalamun ('Anaza domain north-east of Damascus)	Yuzbashi (captain)	450	1	450	5,400
	Lieutenant	200	1	200	2,400
	Soldiers	160	50	8,000	96,000
	Total	—	52	8,650	103,800
Total Damascus Sanjaq		**—**	**108**	**17,600**	**211,200**
Sanjaq of Hawran					
A Nomads of Laj'a under Yuzbashi 'Awda Abu Sulayman		160	25	4,000	48,000
B Same under Shaykh Mutlaq al-Qudays		160	15	2,400	28,800
A Jabal Druze under Hazima's grandson and Ibrahim	Yuzbashi	550	2	1,100	13,200
al-Atrash	Soldiers	150	50	7,500	90,000
B Same under Shibli al-Atrash		160	25	4,000	4,800
Total Sanjaq Hawran		**—**	**117**	**19,000**	**22,900**
Sanjaq of Balqa'					
A Bani Sakhr nomads under Sattam al-Fayiz		150	10	1,500	18,000
B Men to be employed from the start of local registration		150	70	10,000	126,000
Total Sanjaq Balqa'		**—**	**80**	**12,000**	**144,000**
Grand total		**—**	**305**	**48,600**	**583,200**

Source: See text.

supervision and government presence, either through the posting of detachments of horsemen under Turkish officers, or through the co-operation of local chiefs whose duty it was to oversee security and stability in their respective areas. The translation of this document has been kindly rendered by Professor Halil Sahillioglu.[48]

These forces, numbering only 305 officers and men, were spread out in the four districts lying to the north and east of Damascus and adjacent to the pilgrimage route, as well as in southern Lebanon and the Jawlan where there were also elements of population that had to be kept under check, namely the nomadic tribes and the Druzes.[49] In addition, this province had by then assumed an extra importance for the Ottoman Empire. This was due to the settlement of Circassian immigrants who were arriving in Syria after the continuous campaigns on the Russo-Turkish fronts made it difficult for them to attain a settled life there.[50]

However, it is interesting that Sattam is mentioned in the documents as being 'from the shaykhs of the Bani Sahkr tribe' as he was still, during that period, only one of the sons of the shaykh al-mashayikh. His seat was at Zizia and probably he controlled, with his allocation of 80 men (26.2 per cent of the force), the whole area south of al-Zarqa' and north of Qatrani in al-Karak province. The latter does not seem to have had any Ottoman presence at that time but travel books of the period mention clearly that Sattam was recognized as a leader whose word had to be heard.[51] Undoubtedly his influence was based on the strength of the Bani Sakhr because the 80 irregulars in the employment of the Ottoman Empire were probably never enlisted. As was the case in other areas where shaykhs' names are specifically mentioned, Sattam received his 144,000 piastres every year. With the rate of exchange as it was, that amounted to 1,200 gold pounds sterling which would have been an excellent addition to his yearly income. The Ottomans, after re-establishing their governmental administration in al-Salt in 1867, must have found after many bitter experiences that governing through local chiefs was the most successful and least costly method. Transjordan seems to have had greater stability in the northern and al-Balqa' districts during this period than it had ever had since the Ottoman conquest in 1516. Sattam was a wise son of the land who had behind him the might and fame of the Bani Sakhr horsemen.

One final aspect of Sattam's activity that is of note is his establishment during the 1870s of a flour mill at the waters of Amman.[52] Evidently he discovered the need for such a mill after settling in Umm al-'Amad but it may be also that he had plans to take over the fields around Amman, which besides being rich had the great advantage of being next to running water. This is a privilege that al-'Adwan had through their control of the waters at Hisban and it would have been only natural for Sattam to try to attain a similar position. However, he was confronted by the opposition of the Balqawiya tribes who lived around Amman and had already for many generations laid claim to its fields. He decided to be diplomatic and hoped, through building a flour mill which would serve everyone including the Balqawiya households, to allay their fears. In summer Sattam used to transfer his major encampment to Amman and, according to a story heard by his grandson Shaykh 'Akif,[53] was in the habit of having the end of his large *bayt al-sha'ar* (tent) next to the edge of the running water of the Amman spring. Open conflict between the two tribal confederations was averted by the fact that in 1878 the first group of Circassian immigrants arrived in Amman and were allotted by the Ottomans these same fields. Sattam, being in charge of discipline and order in Zizia, was not responsible for the Amman area, but it is interesting to note that the Bani Sakhr became the friends and allies of the newcomers. In 1904 they used their good offices to bring peace between the Circassian community and the Balqawiya tribes in the Harb al-Balqawiya (al-Balqawiya war), which marked the end of a 26-year struggle between the two communities.[54]

Sattam is still remembered as an outstanding leader whose descendants have maintained the mashyakha of the Bani Sakhr tribe since his death. His son, the late Mithqal Pasha al-Fayiz, was the largest landowner in Transjordan with title-deeds in his possession for over 100,000 dunums or 10,000 hectares. He continued the tradition of his illustrious father and is known as the first Jordanian to employ a tractive unit on his estate. His sons, grandsons and great-grandsons continued to engage in agricultural activity of some form or another, thus proving that Sattam's pioneering spirit, resourcefulness and far-sightedness have borne fruit. The whole countryside between Amman and Qatrani has benefited greatly from the originality and stability provided by an outstanding son of the desert who was an organizer and a shaykh in his own right.

11 Circassians and Chichen: Newcomers to the Holy Lands

Visiting the southern parts of Syria in March 1879, Laurence Oliphant, an experienced traveller who had made earlier trips to the Ottoman Empire in 1855, 1860 and 1862, provides interesting information about the first Circassian arrivals in the area.[1] Their head man in Qunaytra, Isma'il Agha, had for six years been held prisoner by the Russians. After visiting the new settlement Oliphant wrote:

> He spoke sadly of his expatriation and the fate of his countrymen, allowed no rest, but ejected in a wholesale manner, first from Russia to European Turkey and now from Bulgaria to Syria. There were altogether about 3000 Circassians in the vicinity who although they had only been there a few months were already establishing themselves in comparative comfort. They were grouped in seven villages all of which they had themselves built and had brought enough property with them to purchase a few cattle so that they were not in absolute want, though some of them were very poor. The Goverment was still supplying scantily the necessities of those who needed it; but it is evident that a government whose resources are not sufficient to buy food for its own army cannot do much to feed scores of thousands of Circassian and Moslem refugees from all parts of European Turkey.[2]

Two weeks later, Oliphant visited Amman and wrote:

> In the ruins of the theatre we were quickly surrounded by a

> group of Circassians who have been settled by the order of the government like those I had met at Qunaitrah. They said that 500 of them had arrived here about three months previously but that the majority had speedily become discontented with their prospects and had gone away. 150 including women and children were all that remained. They had already planted a vegetable garden, had got a good herd of cattle, a flock of sheep and seemed likely to do well.[3]

This information seems to have been absolutely correct, as the Salname of 1302 H/1884 AD clearly states that the number of Circassian immigrants living in the province of al-Balqa' during that year, according to the records of the Kumisiyun al-Muhajirin (Immigrants Bureau),[4] was as follows:

Circassians				*Chichen*	*Rumeli*
Khana	*Persons*	*Males*	*Females*		
43	162	90	72	None	None

This small group arrived in Amman after a year of hardship and continuous suffering. Originally they migrated from the Caucasus in 1864 and were settled by the Ottomans in the Bulgarian city of Plevna and the countryside around it. Fourteen years later they had to migrate again as Turkish rule in Balkan areas was meeting with stiff resistance and defeat. When 'Uthman Pasha, the commander of the Turkish garrison in Plevna, was forced to surrender to the Bulgarians who were supported by the Russians, the Circassians had no choice but to follow the retreating army. Most of them were settled in Anatolia while smaller groups were directed to Syria where they settled around Aleppo, in the Jawlan, Palestine and Transjordan. Information about the numbers involved in this operation during 1878 is not clearly reported but the consul-general of France in Beirut estimated, in a dispatch dated 16 April 1878 to the Minister of Foreign Affairs in Paris, that the number of arrivals was already 25,000 and that the figure might rise to 100,000.[5]

The Transjordanian group of 500 were among 3,000 Circassian emigrants[6] who left Cavalla in eastern Thrace on board the Austrian Lloyd steamer *Sphinx*[7] on 1 March 1878, bound for Latakia on the Syrian coast. At 7.00 a.m. on 5 March the steamer was nearing the rock Klito off Cape St Andreas in Cyprus, intending to go to Famagusta to await better weather and take provisions for the emigrants.[8] At 3.00 p.m. a heavy sea washed away 40 refugee Circassians; at 6.45 smoke was seen issuing from the fore-hatch and a little later the steamer grounded on a sandbank. During the night attempts were made to extinguish fires but without success. Five hundred emigrants perished and on 6 March the surviving emigrants were landed. Two days afterwards a French gunboat[9] received on board the master and crew who fled from the wreck as the Circassians threatened to murder them. Next day HMS *Coquette*[10] took on board the first mate. The vessel was completely destroyed.

A short while afterwards, the survivors were transferred to 'Asqalan or Acre from where some of them were sent to Nablus and later to Amman.[11] These were all from the Shabsugh tribes and lived in and around the Roman amphitheatre. A few years later other groups followed who were from the Qabartai and Bzadugh tribes. The Qabartai lived in a separate quarter, a short distance to the south of the Shabsugh houses, and ever since then Circassian Amman has had a Hayy al-Shabsugh as the northern quarter and a Hayy al-Qabartai as the southern one. The Bzadugh moved to Wadi al-Sir[12] where they founded a new village which is now part of metropolitan Amman starting from the sixth circle.[13] In 1892 a new wave of immigrants arrived via Damascus and they built their homes in the Ras al-'Ayn area nearer to the Amman water springs. The Shabsugh and Qabartai[14] had by then become so established that they called the new quarter Hayy al-Muhajirin, or the Immigrants Quarter, the name that is still used. In the years 1902–5, a small group of Chichen arrived and were settled in Swaylih, half-way between Amman and al-Salt, and in al-Zarqa', 22 kilometres to the north of Amman.[15] The Circassians of Jarash settled there in the year 1891.[16] Other waves of Qabartai and Bzadugh arrived in 1901 and 1909 and settled in al-Rusayfa between Amman and al-Zarqa';[17] in 1901 some of them settled in Na'ur.[18]

In all these settlement operations, supervised closely by the

Ottoman government, there was a common factor of great importance. In every site there was running water which could be used by the new settlers. The communities were all, with the exception of Jarash, in al-Balqa' and within a radius of 25 kilometres from Amman, which became the centre of Circassian and Chichen life in Transjordan.[19] An attempt accurately to record the numbers of those who came during these migrations to Transjordan is not possible as there is no mention of any actual number other than that of the first wave which arrived in 1878; however, other information is collected in table 11.1. It must be stressed that the figures for Circassians are estimates. Dr Waleed Tash, in a paper presented to a seminar on 'Minorities in the Arab East and Israeli attempts at manipulation', held in Amman between 12 and 15 September 1981, stated that the number of Circassians in Amman alone in 1914, according to contemporary individual sources, was about 5,000. Dr Jawad S. Idris, another Circassian from Jordan, stated on 16 September 1978 that the total number of Circassians in Transjordan during 1941 amounted to 19,621. Since this census was taken for the purpose of war rations, it could well be inflated and the real number is probably nearer 15,000.

The motives behind the Turkish plans to settle Circassians in Syria have been discussed by many writers over the years. Some, like G. H. Weightman, thought that they were settled for strategic reasons as well as out of considerations of religious piety and charity.[20] Seteny Shami thought that the overriding concern of the government was with agriculture. Supporting her argument with Karpat's opinion that this was largely due to the loss of the Balkans, the main agricultural region of the Empire, she gave this as one of the reasons for sending the Circassians to the grain-producing areas of Syria. She also agreed with Migdal that the inability of the Ottomans to provide security against the raids of large nomadic and semi-nomadic beduin tribes led to the withdrawal of the peasantry from the plains and valleys. With Ma'oz, she asserted that the Ottomans created a buffer against the dissident sectors of the population by settling Circassians, Kurds, Armenians and Assyrians on the fringes of the Syrian desert.[21] Brawer in turn expressed the opinion that a few

Table 11.1 Date of migrations to Transjordan

Locations	Area	Arrival dates	Estimated number of all arrivals	Actual population in 1930	
				Circassians	Chichen
Amman	50363	End 1878	150		
		1880	500		
		1892	200	1,700	
		1909	40		
		1912	50		
Wadi al-Sir	36556	1880	900	2,000	
Na'ur	46445	1901	300	500	
Swaylih	3743	1905	300	150	400
Al-Zarqa'	4771	1905	150		200
Al-Rusayfa	2352	1902	50		150
		1909	50		
Al-Sukhna	3875	1905	60		100
Jarash	12257	1885	800	1,500	
Total			3,500	5,850	850

Sources: Areas of villages have been listed as given in the annual report of the Lands and Surveys Department in Amman for 1975.

Arrival dates are based on information in the reports of Peake, al-Mufti, Schumacher, Conder, George Adam Smith, Grey Hill, American Expedition to Syria 1899–1900 and the chronicles related by descendants of Circassian immigrants such as General Jalal Khutat and ex-deputy Jamil Shuqum.

Estimated number: Salname of 1302 H/1884 AD for first group in Amman only. All other figures based on published population figures from which estimates were drawn after taking natural population growth into consideration.

Actual population as given in *The Handbook of Palestine and Transjordan*, ed. Harry Charles Luke, chief secretary to the government of Palestine, and Edward Keith-Roach, deputy district commissioner of Jerusalem, in 1930.

thousand Circassians were settled on the fringes of the inhabited lands of eastern Syria and in Transjordan and Palestine to strengthen the security measures against the incursion of aggressive beduin tribes, and to restrain rebellious minorities.[22] Luke and Keith-Roach believed that Sultan 'Abd al-Hamid II shrewdly decided that these wild and warlike additions to his Muslim subjects could be used to the best advantage as a counter-irritant on the Arab marches of his empire.[23] Writing about activities and Ottoman reforms that helped stimulate economic development, C. Gordon Smith mentioned that the populations of cities grew while the depredations of the beduin in the east and south of the country (meaning Palestine) were to some extent controlled by the more forward policy of the Turkish administration, which after 1870 included the planting of Circassian colonies on the desert fringe.[24]

Most probably the Ottomans never had a firm objective in the settlement of the Circassians and the Chichen. The truth of the matter is that they were confronted with a difficult situation which involved people whose settled life had been disrupted, first in the Caucasus and later in the European regions of the Empire, as a result of military defeats on both fronts. The Ottomans felt, both as Muslims and as a state, that they had a moral obligation to assist these people who had fought valiantly in the cause of Islam and the Empire. However, they could not settle them in Turkey because of the animosity shown to the newcomers by Turkish villagers in Anatolia. Besides, the refugees were farmers who need land for their livelihood and the lands in the Jawlan and al-Balqa' were not fully cultivated. The opportunity was there and the newcomers were settled in Syria with the primary aim of engaging them in agriculture. The Circassian role as a buffer between the 'desert and the sown' must have developed after actual settlement took place. The reports about the animosity between the population who were living or tented on the land and the newcomers tend to give one the impression that those who were well-armed and aggressive found it advantageous to ally themselves with the authorities and become soldiers and agents in the area.[25]

As refugee newcomers, who were known for their agricultural and pastoral activities in their former abodes in Bulgaria and the Causasus, the Circassians and the Chichen were naturally looking forward to becoming involved in agriculture in their new homelands. With the exception of the few crafts that their simple life required, agriculture and animal husbandry were the only fields of economic activity open to them. Trade was not particularly interesting for them and it was quite surprising for the people of Amman to see a Circassian, 'Uthman Hasan, open a shop in the market-place and start a commercial activity in the 1930s.[26] Until then, trade was confined to Damascene and Palestinian merchants who, with the growth of the city of Amman by 1900, found it profitable to open shops in it. Jarash furnishes another example of Circassian lack of interest in trade, as it only had Damascene merchants. The other Circassian communities traded with the merchants of Amman due to its proximity to their villages, and do not seem to have found any need to become traders themselves.[27] Circassian interest remained in land. They were good farmers who were willing to work the soil and, with their small savings and the financial help they received from the Ottoman authorities, they were able to make the initial investment required for farming. The Ottomans, for their part, having settled them on these lands, were willing to give them legal as well as military support as they formed agricultural communities in the midst of Transjordanian tribes.

Agricultural activity in al-Balqa' in 1878, when the first Circassians came, was already increasing. The areas that the government allotted to the newcomers, only 10 or 15 kilometres to the north, were all within the domains of settled and semi-settled tribes who were already cultivating fields of wheat, barley and millet. Lands that were not used for cereal production were exploited as pasture-land for the many flocks of sheep and goats al-Balqa' tribesmen seem to have had.[28] People from al-Salt were spread in different localities and were engaged in farming the fields of Amman itself as well as the neighbouring villages.[29] Activity was growing year by year and the supply of arable land seems to have been greater than the demand of the indigenous population; but the tribes were not paying all due land taxes to the governor in al-Salt, and so, by settling new landowners on land already being cultivated, the authorities would be able to recoup a source of badly needed revenue.

The arrival of the new settlers, after the initial period of surprise and despite any sense of duty towards fellow

Muslims, must have been a cause of deep concern for the indigenous population, especially those who owned property in these areas or claimed rights to it. Many groups were directly concerned[30], but those most seriously affected were the clans around Amman. The Haddid clans, who either owned the lands or claimed them, were among the oldest of the Ahl al-Dira (People of the domain).[31] They and their allies, the Da'ja, formed the eastern line of al-Balqa' confederation, in direct confrontation with the Bani Sakhr. The latter were pushing northwards and westwards, and by then had already succeeded under Sattam's able leadership and diplomacy to enlarge their domain of pasture-land and tenting ground. Sattam was greatly interested in the running waters in the area, and building a flour mill in Amman was probably part of his design to take hold of it.[32] Al-Haddid realized the situation clearly and, calling on their allies in al-Balqa' confederation for moral support and assistance in the field whenever necessary,[33] succeeded in checking these advances. This did not mean, however, depriving the Bani Sakhr, or any of the other large tribes such as the Shararat and Ruwala, of the time-revered custom of access to water.[34] The members of al-Haddid, like those of other beduin clans, were reluctant to become involved with government departments or to pay taxes. They did not therefore have any title-deeds and generally did not pay any taxes other than those on animals, known then as widy (beduin usage based on the classical *ada'* meaning that which is rendered). They were not unique in the province of al-Salt at that time. The best impression of land ownership in general is given by Laurence Oliphant, after his visit to the area in March 1879:

> The Turkish Government has not yet ventured to enforce the conscription for the army in al-Salt, and in order to avoid being liable for it, none of the inhabitants have taken out tapoo papers or title-deeds for the real property which they occupy and cultivate. The consequence is that throughout the whole of the Belka there is not an acre owned for which a legal title can be shown. They now hold by prescriptive right alone and numerous quarrels arise in consequence over the possession of land. The hillsides in the immediate vicinity of Salt are covered

> with the finest vineyards, from the grapes of which excellent wine could be made if the art was properly understood; but not one of those who cultivate them can produce a scrap of paper giving him any right to do so in a state of things which at present makes the transference of land, except by the unsafe process of a mutual agreement, impossible. The whole country, in fact, is governed by use and custom, tempered by the somewhat rough principle that might makes rights.[35]

The actual colonization operation for the settlement of the Circassians in Transjordan by the Ottoman authorities was carried out through a policy that was simple and yet consistent. As the government claimed all miri lands, no matter what tribes might tent on them, it felt that it could settle the newcomers without any legal hindrances. When the ownership of the lands allotted to thc Circassians came into question, the generally simple people who staked a claim were caught on the horns of a dilemma.[36] Government agents started an inquiry and asked: 'Who owns the land?' When a clan or person answered 'We do', the devastating question followed: 'Where are your tapu deeds and when did you pay your taxes?' In most cases the back taxes claimed by the government were equal to or even more than the value of the land at current prices. The inhabitants either dropped their claim or did not press it any more. Then followed the actual take-over when the government settled the Circassian colonists on the lands declared vacant, furnished them with seed corn and oxen, and freed them from the payment of taxes and military conscription at the initial stages.[37] Needless to say, there were those who paid their taxes, whether on time or in arrears, and thus retained their title to lands being cultivated by them. The biggest group of these were the Saltiya tribesmen who seemed to have had enough material means at their disposal to meet such financial demands. There were also members of the 'Abbad tribe in Wadi al-Sir and al-'Ajarma in Na'ur who managed to meet the requirements and maintain their ownership. But the majority lost their claims through default on payments. An interesting story is related by Guy Le Strange, who visited Wadi al-Sir during his trip around 1885 and stayed overnight with a camp of the 'Abbad. He wrote:

> Although they were hospitable, our hosts took up the best part of the night detailing their grievances to us and requested our advice on the important point of how £100 might be obtained on loan to rid them of their enemies. It appeared that certain lands belonging from time immemorial to their tribe, for which moreover, they held title-deeds, had been (by government) granted to and were occupied by the immigrant Circassians. We suggested that a petition forwarded with the title-deeds to the government would doubtless set matters right; but in reply we were assured that so doing, unless much bakhshish went with the papers, would only lead to the loss of the deeds without there being the smallest chance of the tribe obtaining any re-establishment in their right. Cheaper than this, they said it would be to bribe the Circassians to decamp and take up their quarters on somebody else's land and for this purpose a hundred pounds were needed which we however deeply regretted being perforce unable to put them in the way of obtaining.[38]

Another example of these differences and disputes is provided by al-Haddid clan of the Balqawiya tribe. They tried to solve their problem differently, through a sale deed which was dated 8 Rabi' al-Thani 1297 H/21 March 1880 AD, well over a year after the arrival of the Circassians in Amman. The document reads as follows.[39]

> Praise be to God alone.
>
> In a majlis [council] this legal transaction of sale was concluded, in the district of Al-Salt and in the presence of those whose names are hereunder mentioned, he who is present in the majlis Salih Effendi al-Nasir Abujaber, of the Christian Community living in al-Salt, bought for him and his sons Farhan and Frayh in equality between them with their own money and not that of others, from the sellers to him in final sale, irrevocably Sayil Ibn 'Abd Allah al-Haddid, Sayah Ibn Idfayl al-Haddid and Ishtaywi Ibn 'Ali al-Haddid present with him in the majlis, representing themselves and acting on behalf of their party Nuwayran Ibn 'Ali al-Miqbil al-Haddid, Thunayan Ibn 'Ali al-Haddid, Khalil Abu Idraywa al-'Abid and Husayn Abu Inshaysh and the sons of Muslim Abu

الحمد لله وحده

بمجلس عقد هذا البيع الشرعي الواقع بقصبة السلط بحضور من تذكر اسماؤهم فيه ادناه اشترى
هذا الحاضر بالمجلس صالح افندي الناصر [illegible] القاطنين بالسلط له ولولديه فرحان
وفريح بالسوية بينهم بمالهم لانفسهم دون مال غيرهم من البائعين منه بيعا باتا وهم صالح بن عبد الله
الحديد وصياح بن اذفيل الحديد و[illegible] بن علي الحديد الحاضرون معه بالمجلس بالاصالة عن انفسهم
وبالوكالة عن جماعتهم وهم [illegible] بن علي المقبل الحديد و[illegible] بن [illegible] الحديد وخليل ابو ادريوح [illegible]
وحسين [illegible] واولاد مسلم ابو عمارة سالم و[illegible] و[illegible] ابو ادريوح [illegible]
[illegible] وقاسم ادريوات العيد ومحمد [illegible] ادريوات
[illegible] عشيرة الحديد بقضاء السلط [illegible]
[illegible] المنازع [illegible]
[illegible] جميع الارض المفتلح الواقعة باراضي قرية عمان المعروفة بارض [illegible]
[illegible] من جهة القبلة الطريق السلطاني [illegible] من جهة الغرب [illegible] الفاصلة بين ارض
[illegible] سيل ماء عمان والقناة ومن الشمال ارض [illegible] جهة القبلة [illegible]
[illegible] بجميع حد وذلك كله وارضه كلها وحقوقها كلها وطرقها [illegible] الشروط المفسدة
[illegible] خاليا عن [illegible] ولا خيانة ولا فساد ولا مواطئة ولا تلجئة بل بيع رغبة [illegible]
[illegible] والقبول [illegible] الصحة واللزوم والتسلم والتسليم
بثمن قدره وبيانه خمسون الف قرش [illegible] بقصبة السلط [illegible]
تاما وافيا [illegible] المرقومين من يد المشتري المذكور
حسب اعترافهم بالقبض قبضا شرعيا [illegible] المذكورين جميع الثمن [illegible]
[illegible] البراءة الشرعية براءة قبض واستيفاء [illegible]
بايجاب شرعي وقبول معتبر مرعي وتسلم وتسليم صحيحين شرعيين [illegible]
هذا جميع ما وقع عليه عقد البيع المذكور [illegible] هؤلاء المذكورين [illegible]
[illegible] هذا العقد بعد صحته وتمامه ونفوذه وابرامه [illegible]
تفرق الابدان وذلك كله [illegible] العاقدين [illegible]
هذا المشتري [illegible]
[illegible] المسمى هذا الكتاب [illegible]
عليهم واقروا [illegible]
[illegible] مكرهين [illegible]
الحال [illegible] تاريخ [illegible] شهر ربيع الثاني سنة سبع [illegible]

[illegible]

شهود الحال

Sale deed of the arable lands in the village of Amman in 1879

Hanada, Salim and Musallam and the sons of Ghadir Abu Idraywa and Sulayman, and Muhammad Shihada al-'Abd, and Hamid al-Anbar al-'Abd and Qasim Idraywat al-'Abd and Hamdan Idraywat al-'Abd and Sulayman al-Kuhaywi and Thalji al-Mansur and 'Awda al-Khala from the tribe of al-Haddid in the Qada' of al-Salt, have sold to him in irrevocable and final sale that which is theirs in equality, their property and in their adverse possession without objector or disputor until the issuance of the final sale deed, all the arable fields falling in the lands of the village of Amman known as the land of al-Haddada bound on the south by the Sultani Road and a hill stretching from the west to the field of 'Ayd Itlas of the Id'aja clan, and on the east the sayl of Amman and on the north the lands of Khunayfasa and the land of al-Haddada and on the west orchards belonging to the Id'aja between the land of Id'aja and the land of al-Haddada and stretching to the south to the mentioned hill, with all its boundaries, its lands, fields, rights, roads and paths and that with which it has been known or to it attached, in accepted rights legally bought and irrevocably sold, with full agreement and consent, permissible by law and free from any hindering conditions, annulling meanings, or weakening words without any fraud, perfidy, corruption, collusion or deferment but a sale of consent and transfer of ownership to a new owner and a serious purchase containing offer and acceptance and the conditions of legality, consent rendering and take-over for a price amounting to fifty thousand piastres in gold and silver coins currently in use in the Province of al-Salt in kind, paid presently as price, complete and final as purchasers complete this. All that, to them received by the hand of the sellers above mentioned from the hand of the said buyers, as per their testimony, to have received it in legal form. The said buyers have been released of all and part of the agreed upon price in the legal context, the release being one of receipt and exaction of amount not a release of dropping and absolving rights.

The contract of this transaction of sale ('aqd) was issued with legal agreement and recognized traditional acceptance and sound and legal delivery and take-over both legally recognized and this said buyer has received what he has signed for in the mentioned contract of sale through delivery to him by those

> mentioned sellers all of that free of every obstacle or contest and they have parted from the majlis of this contract after its conclusion, completion, execution, effectuation, issuance and becoming binding, like bodies parting, all that, after these who have bound themselves affirmed that they have seen it all, knew it, and accepted it, as these sellers are bound, should this buyer be accorded less than his rights or any part thereof to deliver to him all that this transaction of sale named in this book has specified. They have undertaken to do all that, everyone whose name has been written at the end after it has been read to them and they confirmed that they have understood it and taken note of it, whilst in condition of good health and perfect mind, of their own free will and without being forced, being free of indisposition and none of them being ill or otherwise, to prevent the legality of the undertaking or the legitimacy of execution and as the case has been as such written was that which is happening on the 8th passed from the month of Rabi' al-Thani in the year ninety seven two hundred and one thousand.

The document was signed and sealed at a most impressive council in al-Salt. It was witnessed by an illustrious array of 14 leaders and dignitaries of the city.[40] Since it was only discovered in 1964, 85 years after it was written, it was not possible to ascertain the true story of the important document as the witnesses were all dead. However, knowledgeable dignitaries in 1964[41] believed that the title-sale was in all probability concluded in an attempt to avert or forestall the take-over by the government of all the arable lands in the village of Amman. Al-Haddid clansmen evidently felt that, with the arrival of the Circassians, a colonization scheme was being carried out that would inevitably involve more of the lands they considered to be theirs. Therefore they called for help on their friend, the brother of their partner in al-Juwayda, Salih Abujaber. The plan most probably was to have the deed signed, sealed and registered at the seat of government at al-Salt; and then, backed by Salih's influence both in his home town, al-Salt, and in the governmental circle in Damascus, to be accepted there as a valid transaction involving legal sale of land. In other words, the attempt was to try to obtain evidence that the land was owned by Salih who had bought it in a legally accepted form from its previous owners. But there was the other aspect to the affair, namely the taxes. It is not known whether Salih and al-Haddid tried to settle any due taxes, but considering what was written a few years afterwards, it is safe to assume that the sums requested were so high that they decided to cut their losses. This is confirmed by the fact that Salih's descendants did not inherit any property in the district of Amman on his death in 1897, while the descendants of al-Haddid clansmen did not own a single dunum west of the Sayl in the midst of Amman. All the lands mentioned in the documents were in 1936 the property of the Circassian families who were cultivating them.

The failure of such attempts to protect the interests of the population residing in the area led to many disputes and fights. Reports about the period confirm that 'animosity between the local beduins and the Circassians started to build up',[42] as the beduins resented the newcomers, not because they were Circassians and non-Arabs, but because they were just another tribe who, it was feared, would infringe on what the local beduins considered to be their pastoral and water rights. In other words, beduin resentment of the Circassians was not derived from racial or national sentiments and considerations but from purely economic ones. Waleed Tash, a Qabartai Circassian himself,[43] is not alone in this line of thought and is supported by G. H. Weightman, who wrote:

> The land assigned to the early Circassian colonists according to the laws of the Ottoman Empire were mostly government lands which had neither been farmed nor taxed previously. Nevertheless some of the local beduin tribes regarded the land as rightfully theirs and initial enmities developed amongst the new settlers and the local bedu peoples. Although the Ottoman authorities consistently supported the Circassians against their rivals it was clearly to the advantage of the Turks to play the two groups against one another in order to insure their continued dominance.[44]

Although there seems to be full agreement that the newcomers met not only with hardship in the early years after their arrival in Transjordan, but also with stiff resistance from the local beduin, who tented on the land, no mention is made of one basic fact. This is that the Circassians did not arrive as colonists with an organized plan to settle, if necessary by force, but as refugees who did not have anywhere else to go after the Ottoman authorities assigned to them the different areas to dwell in. The early colonists approached their new home in the 'Holy Land' with religious awe and great reverence. However, the economic factors could not be ignored and differences with the local population were inevitable; in many cases they were hard and bloody. Killings took place on both sides and the Ottoman government was perhaps unable to prevent this, but, more likely, intended to allow the differences to grow and persist. The government's purpose may have been to deepen the rift between the two communities and maintain the wedge that was placed east of the settled areas against the local beduin and nomadic tribes alike.[45] Otherwise, how can it be explained that in many instances fights broke out less than 20 kilometres from al-Salt, the seat of Turkish authority, even when its dominance over the area had, by the end of the nineteenth century, become fully operational?[46] This phase of turmoil did not, however, last for more than 25 years.[47] Matters seem to have been settled completely in favour of the Circassians in 1904 after a battle, Harb al-Balqawiya, in which al-Haddid clans and their allies were forced to accept the new facts.[48] Against them were the Circassians, the Bani Sakhr, who by then had concluded an alliance with the newcomers, and the formidable forces of the Ottoman administration at al-Salt and Damascus. After that, life for the Circassians in al-Balqa' acquired a higher degree of stability, and relations with their neighbours became more normal and friendly. The alliance[49] during these 25 years between them and the Bani Sakhr, the inveterate enemies of the Balqa' confederation, was undoubtedly strengthened by their leaders having a great interest in the development of agriculture. Sattam and his successors were already on friendly terms with the Ottomans, having agreed to become governors of Nahiya al-Giza. They supported the Circassians with a double aim of keeping the Ottomans happy and annoying their enemies al-Balaqawiya. The Bani Sakhr continued to reap the benefits of this policy until the end of Ottoman rule in 1918.

Life in Amman and its neighbouring settlements was simple and ran at a slow pace. The newcomers were generally farmers but among them were carpenters, blacksmiths, goldsmiths, silversmiths, leather and dagger craftsmen, and carriage makers.[50] Most of their tools, equipment and utensils were made of wood[51] which was abundant in the forests that surrounded their new homeland. It is said that they were the first to have brought tea into Transjordan,[52] and it is nearly certain that their ox-carts were the first wheeled carriages to be driven on the rough roads they prepared in Transjordan during modern times.[53] Crops produced by them were those already known in Transjordan and it is on record that the Circassians did not introduce any new products. Co-operation among them was at a high level and it was customary for the whole village to assist a newcomer in building a house. Their society was settled, with its deeply entrenched traditions, and their standard of living probably quite high, as within a few years of their arrival they had three flour mills in Amman alone,[54] while all along the Wadi al-Sir were mills, sometimes with modern houses and gardens nearby.[55]

Farming was maintained on a personal or household level. Chronicles relate that the early arrivals were allotted lands by Ottoman government officials and that since their numbers were still limited there was no difficulty concerning fields, and every family obtained the area of cultivable land it required. During the 20 years that followed, however, numbers of both those already settled on the land and newly arriving families increased. The land at the disposal of the government for settlement purposes was therefore no longer sufficient, and a plan of redistribution was resorted to, probably in 1308 H/1890 AD.

A Circassian dignitary, Mr ‘Umar Tihbsum,[56] whose father Musa, originally from Qwatz, West of Nalshik, came to Transjordan in 1305 H/1887 AD when he was only 13 years old, related interesting information about the distribution of land. He confirmed on 26 June 1985 that his father told him that the first allotments were made in Amman around 1890 when Mirza Pasha – then a commander in the police force and

A Circassian straw cart that was pulled by a pair of oxen and of great service during harvest (courtesy of Dr G. Van der Kooij, Rijkuniversitat, Leiden, 1983)

the leader of the Transjordanian Circassians during the first three decades of the twentieth century[57] – came to Amman with instructions to that effect. The instructions were issued by his commanding officer, General Khusru Pasha, commandant of police in the wilaya of Syria. The allocation of cultivable land took place on the basis of three grades: the smallest being 60 dunums; the medium being 90 dunums; and the largest being 120 dunums.[58] A tapu document (title-deed) was then issued to every beneficiary by the Tapu Bureau in al-Salt. Fortunately, one of these title-deeds was found among the documents presented recently by the Mirza family to the Department of Libraries, Documentation and National Archives in Amman. It reads as follows:

File no.:	131
Daftar:	Aylul 1309 (August 1891)
Liwa':	Hawran
Qada':	al-Salt

Quarter: Qabartai
Village: Amman
Location: Saqra[59]
Type: Agricultural land
Category: Miri
Boundaries:
South: land of Sayl [running water]
East: land of Ghiwwar[60]
North: land of Khayran
South: land of Zakariya
Area: 20 dunum 'atiq
Earlier owner:[61]
Deed issued by: Immigration Bureau
Issued free of charge for improvement
Holder: Mirza son of Baqa by Ottoman gift
Value:
Amount paid:
Given in return for A'shar Shar'iya [legal yearly tithe][62]
Issued on 5 Shawwal 1312 [2 March 1895]

Unfortunately it has not yet been possible to find any document that gives a list of the families that benefited from this operation in Amman or other Circassian settlements. 'Umar Musa Tihbsum confirmed that, as far as he could remember, the Qabartai of Amman had nearly 100 faddans, the Shabsugh between 60 and 70 and the Muhajirin also between 60 and 70: a total of about 230 faddans or an area of about 21,000 dunums. He was certain that even at an early stage the Circassians started leasing or buying land, since the miri lands, placed by the Ottomans at the disposal of the newcomers, were not sufficient for their agricultural activities or the living requirements of about 1,000 people.

In the other two large settlements, the mayor of Wadi al-Sir, Mr Husni Shuman Sawbar,[63] confirmed that the Circassians in that area in the 1900s were running nearly 250 faddans, while Mr Ishaq Yusuf Khamash[64] confirmed that at a later time the Circassians of Na'ur were running some 85 faddans.

For the Chichen, on the other hand, who arrived at a later stage, there is extant a document which outlines the details of land distribution among them, and the number of khanat that benefited from it. It makes special mention of the irrigated lands in al-Zarqa', al-Rusayfa and al-Sukhna which were Chichen villages. Irrigated lands were definitely cultivated in Amman, Wadi al-Sir, Na'ur and Jarash, but are not mentioned even once by the Circassians who, contrary to the Chichen, seem to have attached far greater importance to the dry farming of cereals on the plateau. The document was prepared at the orders of the wilaya of Suriya on 3 Safar 1323 H/10 April 1905 AD. In it the lands in the villages of al-Haddid (al-Zarqa') and al-Sukhna are distributed to the Chichen and Lazqi refugees from Daghistan. The distribution of the total of 14,400 dunums is shown in table 11.2.

The irrigated areas of the distributed lands amounted to 947 dunums at al-Zarqa' and 1,132 dunums at al-Sukhna or a total of 2,078 dunums. This meant an average of between 10 and 15 dunums per family.

It is to be noted that the system of distribution must have been the same in all the Circassian and Chichen settlements, although there could have been variation in the areas allotted to different khanas here and there. For example, areas in al-Zarqa' were ten dunums larger for every category and this may be explained by the fact that dry farming lands in Amman, Wadi al-Sir and Na'ur are, on account of higher averages of annual rainfall, more fertile and productive than the lands in al-Zarqa'.

The arrival of the Circassians and the Chichen in Syria in general and in Transjordan in particular may have caused some hardship to the population who were already on the land. There were certainly initial differences and fights and

Table 11.2 The distribution of the lands in the villages of al-Haddid and al-Sukhna

Khana	Share per Khana	Total land distributed
78	70 dunums	5,460 dunums
66	100 dunums	6,600 dunums
18	130 dunums	2,340 dunums
162	**88.88 dunums (average)**	**14,400 dunums**

the period of adaptation and assimilation extended over a few decades. However, it is only fair to mention that the contribution of the newcomers was a valuable one, enhancing the quality of life in the areas they colonized. It is true that they did not introduce any basic changes in the system of agriculture, or any new products to the land; but they brought in new farming techniques and were able, through perseverance and hard work, to improve the productivity of the agricultural system as a whole. Through living in Amman and its environs they have acquired, since the establishment of modern Jordan, a good deal of importance. The numbers have been estimated recently at about 60,000 Circassians and between 8,000 and 9,000 Chichen.[65] Their venture into Transjordan has been, after the early hard days, a rewarding one, and their life in their new homes, during the last hundred years, has been indeed an enriching experience for the society as a whole.

12 Madaba and its Christian Tribesmen

Madaba is an old site on the top of a hill 30 kilometres south of Amman (see map 12.1). It is one of the most ancient and important cities of the area, together with Dhiban and Hisban. Archeological excavations have dated an early Iron Age tomb discovered in Madaba in the 1950s to the thirteenth century BC, which is in line with its mention in the Bible (Numbers 21:30).[1] Its name occurs in the Moabite Stone inscription, sometimes called Misha' Stele, discovered in Dhiban in 1868 and presently in the Louvre museum in France. This inscription is dated by archeologists as the ninth century BC and it gives an unhappy picture of turmoil in southern Transjordan when Madaba was occupied for 40 years.[2] Centuries later, during Roman times, it was a prosperous provincial town and continued to be important during early Christian times when it became an episcopal see. The many temples and public buildings erected during these two eras contained the mosaic art which made Madaba and its environs famous. Beautiful mosaics created by its talented and well-informed artists during the fifth, sixth, and early seventh centuries AD are many, but the best is the famous Map of Madaba with its relatively correct geographical rendering of Transjordan, the Dead Sea, the River Jordan, and Jerusalem.

Like many of the other parts of Syria, Transjordan suffered during the Persian invasion in 614 AD, and Madaba was probably destroyed. It was occupied for a short time after the Arab Conquest, but subsequently seems to have been abandoned.[3] This state of neglect continued until the nineteenth

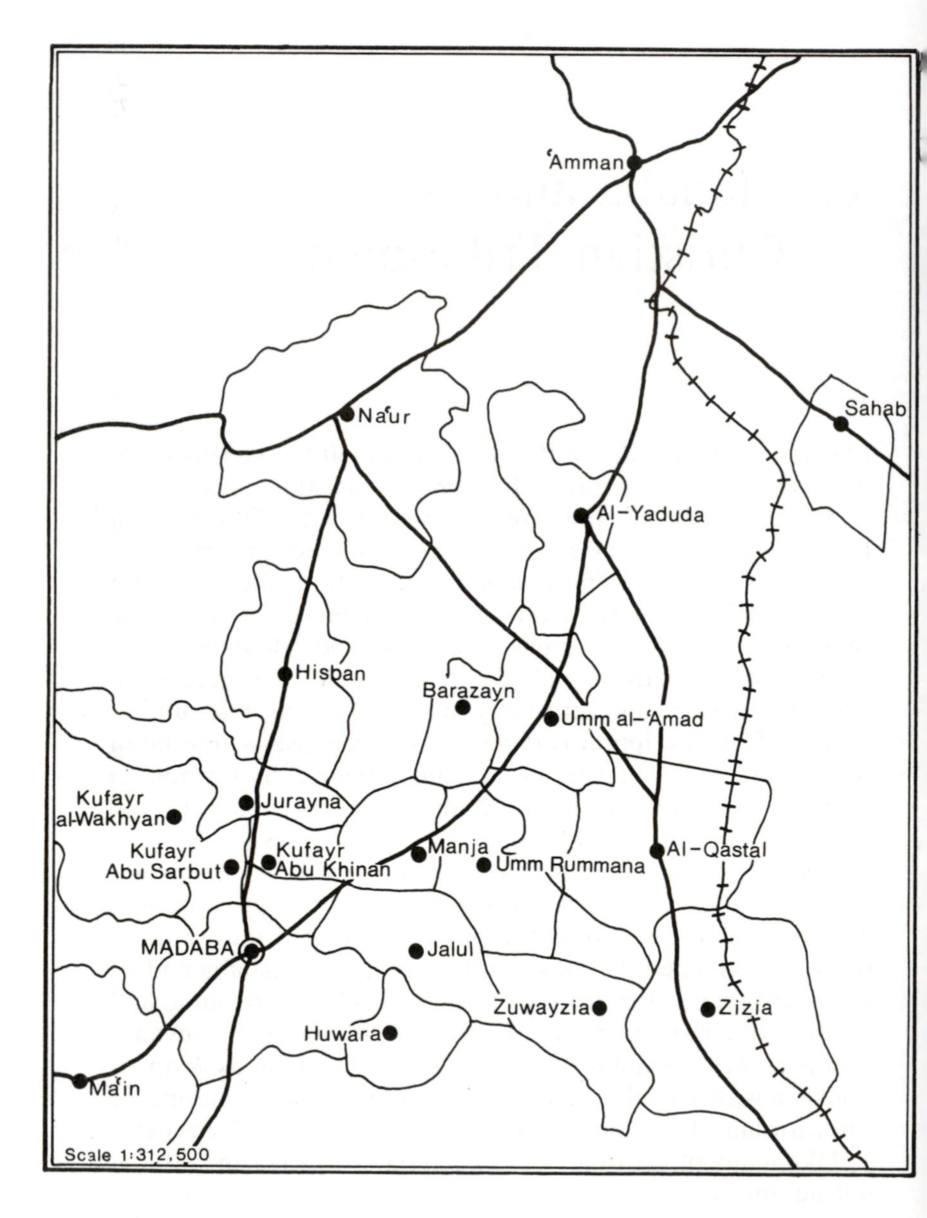

Map 12.1 Situation of Madaba

century when it was visited by Europeans for the first time in modern times. Neither Seetzen (1806) nor Burckhardt (1812) found anything other than the ruins to write about.[4] In 1872 Tristram visited the ruins, which were in a relatively good state of preservation, and thought them to have been a city of great importance. He further reported that part of the country around Madaba was under cultivation where ‘Ayd al-Rudayni was ‘one of the very few Bani Sakhr of high degree who turned his attention to agriculture; cultivating the rich soil of the ancient bank and the neighbourhood of the old city by his slaves; and claiming through nomad descent, personal and not tribal possession of the land’.[5] It is no wonder therefore that the colonization of Madaba in 1880 by three tribes from al-Karak provoked a certain amount of controversy.

These three tribes, al-‘Uzayzat, al-Ma‘aya‘a and al-Karadisha, lived together with few other tribes in al-Karak until 1879. The origins of these tribes are, like so many genealogies among the Arabs, remembered only through chronicles and family stories. Al-‘Uzayzat and al-Ma‘aya‘a are said to be descended from the old tribes such as the Ghassanids, while al-Karadisha is said to have moved from the village of Salkhad in Jabal al-Druz to al-Karak in the seventeenth century. Stories about the origin of al-‘Uzayzat, the largest tribe, are however varied but there are two that are especially prominent in their chronicles. The first is that they were Ghassanids, from Nawa, a village in Hawran;[6] and the second is that they originally came from the area north of Mosul, where they worshipped the goddess al-‘Uzza.[7] The latter story gains credence from the fact that the word ‘Uzayzat is a derivative of al-‘Uzza, the Arabic name of the goddess Aphrodite-Venus, and the fact that this deity was worshipped on a large scale by the Arabs of Petra, the Nabateans, during the fourth century.[8] The story relates that two brothers of this clan seem to have been in Mu‘ta, south of al-Karak, in August 629 AD when the Prophet Muhammad sent the first Arab expeditionary force into Transjordan. The Muslims, 3,000 in all, were defeated by the Byzantine forces and their allies and had to withdraw to the east. The chronicles relate that during these difficult days, the two brothers helped the Muslim army. One of them adopted Islam while the other, the ancestor of the present 'Uzayzat, remained a Christian. Their deed, it is said, was very favourably viewed by the Prophet, who ordered that neither they nor any of their descendants should pay taxes. The 'Uzayzat paid taxes only after the Ottoman governor of al-Karak rescinded this tradition in 1911, immediately after the popular revolt.[9]

In 1877 al-'Uzayzat converted to the Latin faith.[10] This came about when a few of their heads of family rode to Jerusalem where they asked the Greek Orthodox patriarch to appoint one of them as another Orthodox priest.[11] The patriarch evidently felt that there was no real need for a second priest alongside Father Aphramius, who belonged to al-Halasa, the largest Christian tribe in al-Karak. Al-'Uzayzat did not want a clergyman from another tribe and insisted that one of them should be ordained. Determined as they were, they approached the Latin patriarch[12] with the request that in return for their adopting Catholicism, he would appoint for them a priest; the patriarch agreed and sent them Father Alexander and an assistant called Father Bulus.[13]

Together with the other Muslim and Christian tribes among whom they lived, the three tribes were composed of three groups: farmers, sheep and goat owners, and merchants. The whole town was under the leadership of Shaykh Muhammad al-Majali. On 6 November 1879, a dispute arose between the Christian 'Uzayzat and al-Sarayra, a Muslim tribe in al-Karak.[14] Fights broke out and the dispute[15] was aggravated by the fact that, while al-'Uzayzat belonged to the Gharaba (Western alliance) headed by al-Majali, al-Sarayra belonged to the Sharaqa (Eastern alliance) headed by al-Tarawna.[16]

The 'Uzayzat decided to leave al-Karak so as to have more freedom in avenging the insult they had received. Such extremes of action, which may be viewed with surprise nowadays, were considered to be customary and proper at the time. It was possibly because al-'Uzayzat had such an ancient noble heritage, and because great valour and bravery was attributed to them, that they took such a drastic measure.[17] Another cause for their exodus could have been the state of weak security in al-Karak which did not facilitate the development and prosperity of a community like theirs for whom political and economic stability was essential. Effective order in al-Karak was only imposed by the Ottomans in 1894 when Husayn Hilmi Pasha entered the city at the head of Ottoman troops and brought the countryside under control.[18]

Al-'Uzayzat moved northwards and for two years lived among the Bani Hamida, where they practised agriculture and looked after flocks of sheep and goats. Their temporary *dira* (home) was not far away from al-Karak; the young among them ventured there in raids against their enemies.

During this time their affairs were being attended to by the Latin patriarch in Jerusalem as well as by the French consul in Beirut, both of whom felt it their duty to help Latin Christians like al-'Uzayzat in their hour of distress. Probably they also had other reasons: the patriarch wanted to impress on the Orthodox Christians how influential he could be, while the consul felt that this was a good opportunity for France to achieve a presence in Transjordan, a politically virgin area.

Then, in the winter of 1880 al-'Uzayzat, who were joined by the other two tribes, decided to move. A mention has been found to explain why the three tribes chose to settle – of all the available places in al-Balqa' – in Madaba.[19] It should be remembered that the area was sparsely populated and that many lands, suitable for dry farming, were not being cultivated. Amman, already colonized by the Circassians, provides a good example of land that was being cultivated, as do al-Yaduda and Umm al-'Amad. There were probably three main reasons for the choice of Madaba. First, the quality of the soil in the neighbourhood was good and there was a good annual average rainfall. The fields were good farm land promising a good crop. Secondly, there were no people at all living in it. It is true that some of the fields were being cultivated by al-Rudayni of the Bani Sakhr, but these could not have been more than a fraction of the 27, 650 dunums that were around. Thirdly, the ruins provided a good supply of stones for use in building and there were caves, wells, cisterns and pools that could be used with little effort for cleaning and maintenance. Water storage was of vital importance for a settlement that was at least 12 kilometres from the nearest spring at Hisban or Ma'in.

The patriarch wrote to the French consul in Beirut and also to the wali of Syria, the famous Midhat Pasha,[20] seeking their assistance in settling al'Uzayzat in Madaba. By the beginning of April 1880, Father Alexander was determined to force the government's hand and wrote to the Patriarch Vincenzo Bracco saying that, 'A hundred ploughmen have already been ploughing the fields of Madaba for 15 days. The tribes that are determined to settle are three in number, with over a hundred families; they wish to rid themselves of the general conditions of instability prevailing in al-Karak.'[21] The three tribes were al-'Uzayzat (62 men), al-Ma'aya'a (45 men), and al-Karadisha (44 men). The pressure continued with messages to the governor in Nablus,[22] who in turn contacted his superior, the wali in Damascus.

This concerted action met with success at a time when a weak Ottoman Empire wanted to appease Europeans in general and also was bent on settling agriculturalists on the border lands in the hope of collecting taxes from them in due course. It was in line with the settlement of the Circassians in Amman and Wadi al-Sir to the north. There seems to be little doubt that the governor of al-Salt, who was directly involved in these events, must have felt happy that another 150 men, who were willing to defend themselves and their fields, were now planting themselves in the midst of the Bani Sakhr. Midhat Pasha wrote to Patriarch Bracco towards the beginning of April, in reply to his letter of 14 March 1880, assuring him that he had given instructions to the administration in Nablus to facilitate the settlement operation.[23] Furthermore, he confirmed that he had asked the governor to give special consideration to the person whom the patriarch wished to appoint as a religious leader for the newcomers. Simultaneously, the administration in Damascus sent these instructions, on 29 March, to Nablus.

To the Governor of Nablus

We have received your correspondence dated 7 Rabi' al-Awal, 1297/18 February 1880, no. 301, seeking permission to settle the Christians of the Latin faith, who have immigrated from al-Karak, in Khirba Madaba, which is empty of inhabitants, in the Qada' al-Salt; and since we have also received a telegram from the religious head of the said community in this regard, it

> has been transferred to the council of the wilaya's administration. The benefit thereof arrived by declaration notifying that if in fact the said village is uninhabited (kharab) and has no farmers who claim the right of prescription, then there is no objection to giving it to them for its rehabilitation. It is submitted that a few Christian families of al-Karak came and were rehabilitated in Qada' al-Salt. After harvesting their crops they returned to al-Karak and since this is certain and if now these newcomers are settled, to prevent them from migrating again, strong guarantees and undertakings are to be taken from them. Whereupon the necessary is to be effected in accordance with this notice from al-Majlis.
>
> Issued on 17 Rabi' al-Akhir 1297 coinciding with 29 March 1880.

The Ottoman interest was continuous concerning the maintenance of farmers on the land; it was also manifested by their attitude to the Christian farmers of al-Husn in 1868. The collection of taxes, naturally, was the aim of the administration and so they attempted to safeguard the settled population they hoped to tax (see chapter 9).

In spite of these explicit instructions the mutasarrif of Nablus was still adamant in efforts to determine whether the village being settled was really free from any who might claim the right of prescription (i.e. claims of ownership, in the legal language of the day, *haqq al-qarar*).

Evidently he was worried about the Bani Sakhr and, considering that he had many cases of disputes on his hands as a result of the settlement of the Circassians in Amman during 1879, this is quite understandable. He therefore wrote to the qa'immaqam of al-Salt on 17 Rabi' al-Akhir 1297/29 March 1880 saying:

> By reading the text of the answer-order, a copy of which is herewith enclosed, you will understand the situation and that, as well as the note of the Mamuria al-Daftar al-Khaqani[24] stating that there are no farmers or people who claim the right of prescription in the said village (Madaba), it is also necessary to inquire from you about this matter.

Therefore if there are no people or farmers who claim the right of prescription for the said village, work and action are to be effected in accordance with the said order of the wilaya, with presentation of a report about results.

As to the Karkiya who complained of misdeeds of some of the beduin, and infringement of justice – if it is true it is irregular and not permissible. Action is to be taken to prevent misdeeds; there is to be protection from encroachments, and impositions and efforts are to be continuous since the matter falls under the obligations of the local government which is entirely responsible for it.[25]

However, the fact that the governors of Damascus, Nablus and al-Salt gave their consent to the colonization of Madaba did not really make things smooth or final for the Christian tribesmen of Madaba. When they were harvesting the first crop of sorghum in August 1879, ʻAyd al-Rudayni of the Bani Sakhr, who claimed the ownership of the Madaba lands, asked for his share of the crop.[26] The tribesmen refused his request and when the dispute was brought to the knowledge of Shaykh Sattam Ibn Fayiz, he gave him the village of Dulayla to compensate for the loss of Madaba. Sattam waited until the new settlers had started the second harvest in August 1880. Then he visited the village and declared that he wanted a camel load each of wheat, barley and sorghum for every faddan,[27] and that he was asking so little only in the spirit of friendship. The spokesman for the Madaba settlers explained to him that they had undertaken to pay taxes to the government in full and that they were not willing to pay double. As the arguments became more heated Sattam, who was not accompanied by many of his men, left in anger, without partaking of the lunch that was offered him. A few days later, he arrived with a strong retinue, threatened the settlers and broke the *narghila* (hubble-bubble) in the guest-house as a sign of his contempt. Father Bulus was informed of this incident and he immediately rode and lodged a strong claim with the mutasarrif of Nablus and won the case for the tribesmen. The mutasarrif sent a detachment of cavalry and placed it under the comand of Shaykh ʻAli al-Dhiyab of the ʻAdwan, with the request that he seize Sattam and send him under escort to Nablus. This commission suited ‘Ali well since he was probably very angry over Sattam's recent divorce of his sister ‘Alya. He therefore occupied Umm al-‘Amad for three days but failed to capture Sattam who was with his encampment deep in the desert.[28]

Immediately afterwards, the qa'immaqam of al-Salt went out on his tour of tax-gathering and encamped at al-Humar, 12 kilometres to the west of Amman.[29] Beduin shaykhs were in the habit of answering the call of the qa'immaqam when everyone received a present and was notified of the tax amount levied on his tribe's livestock. When Sattam came, he was taken prisoner and sent to Nablus where he was detained for three months. Through a good friend, al-Shaka‘a[30] of Nablus, Sattam sent a letter to the Latin patriarch, who pleaded with the mutasarrif and succeeded in liberating Sattam after he had promised not to molest the Madaba people thenceforth. On his release, the patriarch entertained him for a week in Jerusalem and presented him with 300 gold pounds as compensation.[31] True to his promise, Sattam never entered Madaba after this episode, but continued to be a good friend of its people until his death in 1890. Although Madaba's troubles with its neighbours did not cease after the payment to Sattam and fights and disputes continued well into the twentieth century, no one contested their right to the land.[32]

As well as the outside dangers, life in the town itself was not without its difficulties, and many fights and feuds developed among the three tribes and also among the clans of each tribe. Evidently these were so frequent that their leaders had to develop a system which would provide an effective way of dealing with such controversies. A traveller of the time records:[33]

> The Bedawin in Madaba were collected from all parts and many tribes. One family never made common cause with another unless the safety of its own was threatened. Hence the more convenient way for them of settling a feud was by a pecuniary consideration, though in this case it was much lower than among the best Bedawin tribes. A fixed tariff was instituted to regulate the scale of payment for limbs as well as lives.

				£	s	d
Finger	10	Medgediehs	about	1	13	4
Left hand or arm	50	"	about	7	16	8
Right arm or life	300	"	about	47	00	0

Although this sounds an original way to treat instability in a continuously threatened society, it is doubtful that its application could have been easy. Unfortunately there remain no eyewitnesses who might have attested to its effectiveness.

Life among the population of Madaba was quite different from that among their neighbours. They were all Christians and the influence of the two churches, the Orthodox and the Latin, was very much felt. The Latins had Italian priests who were in direct contact with the patriarch in Jerusalem, who was also Italian and always prepared to seek assistance from the French representatives in Jerusalem, Beirut and Damascus. Not all of the 'Uzayzat became Latins, and the house of their shaykhs, al-Farah, for nearly a century continued to be followers of the Orthodox church. These shaykhs were the descendants of a Greek monk from Crete by the name of Kharalambos, which means joy (*farah* in Arabic). While in al-Karak, he left the priesthood, married Nasra, daughter of Sulayman al-Sawalha, and became a member of the tribe.[34] His oldest son, Hanna, probably born just after 1860, became the shaykh of Madaba and for over 50 years was the leading and most renowned horseman and marksman in al-Balqa'. His nephew Ishaq became shaykh after him and Ishaq's son, Samih, until recently the mayor of the modern town, has become the headman.[35]

The free and even active participation of women in social and tribal life was more pronounced in Madaba than among its neighbouring encampments and villages. As a Christian community it was more tolerant of women in its society as a whole. There was much more social contact with women through gatherings at religious services in churches, at weddings and in social activities. Education in general was better, probably because of the presence of Italian priests, than in the area as a whole, and this improvement was particularly visible among Madaba girls compared with girls of the neighbouring settled population.

Generosity in the new settlement was a very strong element in the codes of honour and tradition. Arabs in general are famous for hospitality, but as in the Abujaber case, hospitality among the Christians of Madaba acquired a personal touch.[36] Notable in this respect were two houses of the ‘Uzayzat tribe, the Shwayhat and Ibn Farah. They became renowned all over the area, especially among the tribes of the dira. As a result they were very favourably mentioned in the chronicles of those days. One account even tells us of a certain generosity competition between these two famous clans and the Abujabers of al-Yaduda. Most of those among the settled population who raised flocks of sheep and goats went out with their flocks for around three months during spring into the badiya for pasture. It is said that the competition happened during one of these beautiful seasons in a wadi some 50 kilometres south of Amman. The incident is further attested by the poet of the Shararat tribe, called al-Tuqayj, in a poem probably written around 1900. Understandably, he gave no verdict as to who of the three was most hospitable, but he immortalized their names through his famous verse:

Thalat biyut mibnayat ‘ala al-darb
Abu Jabir wa Abu Shwayha wa Hanna
Mutkat finin ‘idhum jabihit al-harb
Yi‘julu lil-dayf balikra ma yuwana

which freely translated means:

Three encampments on the roadside
Abujaber, Abu Shwayha and Hanna
United as if in line of battle
They hasten to present food to their guests without delay[37]

The members of the different clans, after the initial difficulties, settled down rather well in their new surroundings. The area of the Madaba fields, perhaps then around 20,000 dunums,[38] was divided between the three tribes. Probably at the start of colonization Madaba had about 110 faddans divided as follows:

Al-'Uzayzat	50 faddans
Al-Karadisha	30 faddans
Al-Ma'aya'a	30 faddans
Total	110 faddans

The younger men of the tribes were the actual labour force. With time, they started cultivating land on a partnership basis with both the Bani Sakhr to the south, east and north, and the Balqawiya, especially the Abu Wandi clans, to the west. For this more widespread activity they started hiring labour who came mainly from the Hebron countryside, the Ta'amira tribe and Jabal Nablus. By the 1900s, the area of faddans had risen to:[39]

Al-'Uzayzat	100 faddans
Al-Karadisha	100 faddans
al-Ma'aya'a	80 faddans
Total	280 faddans

Land was also purchased and rented. Some 15 years after their settlement in Madaba, the well-known clan of al-Tiwal bought lands in al-Zabayir from the descendants of Sattam Ibn Fayiz. They thus became partners in the fields of the 5,764 dunum village and since then the village has become known as Zabayir al-Tiwal. Probably on account of their original differences with Sattam, the settlers' activity did not spread to Umm al-'Amad, which was barely 10 kilometres to the north of their village. This has meant that, in spite of the warm social relations they had with the Abujabers on account of religious affinity and agriculture, few tribesmen from Madaba have worked at al-Yaduda.[40] Generally they themselves became employers a few years after their arrival and farmed with a good number of the smaller clans around them who had fields for cultivation on a partnership basis. With time, they started to buy additional fields from their neighbours and the area of their holding therefore gradually increased.

The economic structure of this new settlement, like that of al-Yaduda, Sattam's Umm al-'Amad and Amman, was different from those of their neighbours. Agriculture became more intensive and the returns were much higher. As well as the cultivation of cereals, these communities were especially active in the field of animal husbandry, both to satisfy their need for meat, milk and butter, and as an extra source of income. Flocks of sheep and goats were kept in every household and it is quite possible that by the time the tribesmen numbered 400 families, Madaba had somewhere between 20,000 and 25,000 sheep and goats.[41] In addition, every household had the absolutely necessary faddan, or pair of oxen, a camel or two, a few donkeys, a mare or a mule, and some cows. The milk produced was consumed as milk, laban (sour milk) or cheese and the surplus processed into butter and then turned into samna (melted butter or Syrian butter ghee). The flocks, in good rainfall years, were sent out to pasture in the eastern ranges, sometimes as far as 50 kilometres away from Madaba; the stay there would stretch between February and early June, when they would be brought back to graze behind the harvest labour in the newly harvested fields. Hay was therefore only needed during the harsh winter months of November, December, and January. Lack of adequate grazing in bad rainfall years would always create a great problem for flock owners, and they might lose as much as 50 per cent of their flocks in two seasons.

Although there are no records or documents of any form about these aspects of economic activity, a reconstruction is attempted below of revenue figures for households that had an average of one faddan of cereal production together with 150 ewes and goats in around 1900.

	Tons	*Ottoman gold pounds*
Agriculture		
180 faddans at an average production of 3,000 sa's of wheat and other cereals	4,980	
Less: Seeds 200 sa's/faddan	(336)	
Share of labour one-quarter	(1,245)	
General expenses	(150)	
Provisions	(280)	
'Ushr (crop tax)	(625)	
Net crop	2,344	
At 3 piastres a sa' or £5 a ton		11,720[42]

	Ottoman gold pounds
Animal husbandry	
25,000 ewes and goats giving a net product of 45 piastres each, on average	11,250[43]
Income from sale of mares, 20 per year	1,200[44]
Grand total	22,970

Taking into consideration that the 180 faddans were most probably owned by 250 households, the average of income per household turns out to be nearly 92 Ottoman gold pounds.

If a harath was making between 10 and 20 pounds per year, it is obvious that the standard of living in Madaba was really much higher than that prevailing in the neighbouring villages. Its people could afford a better standard of living and also save enough to reinvest in agricultural improvement and purchase of new lands from their neighbours. They could spend more on education, and within a period of 25 years a new generation emerged that could not only read and write but could also start working as government employees, teachers, and merchants. Their wealth showed itself clearly when a group of the Madaba tribesmen, headed by Hanna Ibn Farah, in 1915 bid for and won the iltizam (tax farm) of the whole Balqa' district from the Ottoman government.[45] No wonder, then, that another poet of the Ghunaymat tribe, Minawir Husayn, could not contain his admiration for the progress Madaba had made after the turn of the century, when he exclaimed

Ya Madaba

Centuries after your solemn ruin
You emerged to be a lover's dream
Surrounded by encampments so numerous
And generous like a pouring stream
Their gallant league from Karak alighted
Supporters of the distressed in dreary years.[46]

13 Sahab: a Frontier Settlement of Egyptians

The movement of people from Egypt to Palestine and Transjordan has taken place from time to time: generally the Egyptians moved eastwards and northwards across Sinai in an attempt to find employment for their growing numbers. In modern times, the fact that the arable land on both sides of the river Nile became densely populated forced a good number of them to seek work and a new life elsewhere.

At the beginning of the nineteenth century, hundreds of young men crossed into Bilad al-Sham to avoid conscription into Muhammad 'Ali's army.[1] In 1840–1, a few thousand of Ibrahim Pasha's army chose or were obliged to stay behind during the Egyptian withdrawal from Syria.[2] It is practically impossible to find a single area in Palestine and Transjordan that does not have families carrying Egyptian names. The most common among these are al-Masri or its plural form al-Masarwa (from Misr, or Egypt); al-Bilbaysi (from Bilbays); al-Sa'idi (from al-Sa'id, or upper Egypt); al-Tantawi (from Tanta); al-Fayyumi (from al-Fayyum); al-Sharqawi (from al-Sharqiyya province) and al-Dasuqi (from Dasuq). Many families of Christian Copts have kept the name of al-Qubti, given to them when their grandfathers first came from Egypt.[3]

One important migration to Transjordan was that which took place in 1868 when the Suez Canal was being built. To complete its construction, the Egyptian government forced Egyptian farmers into *al-sukhra* (forced unpaid labour). Sensible of the danger that might befall them, the farmers in al-Zaqaziq, a town in eastern Egypt to the west of Isma'iliya, delegated their headman, Shaykh Muhammad Abu Zayd, to handle the situation. He travelled to Cairo where his mission was to seek exemption from forced labour on the grounds that the whole community at al-Zaqaziq was engaged in the reclamation of desert land. This argument, it was thought, would appeal to the authorities, especially since the farmers were about to start paying taxes on crops from the new fields. The authorities, however, were not in a frame of mind to listen to such pleas, let alone accommodate them. They had their hands full with financial problems created by the Khedive Isma'il. Shaykh Muhhamad was therefore told that he was to arrange, without delay, the movement of the Zaqaziq villagers to the Canal Zone and that he was expected to be at the head of the group and to set a good example by abiding by the wish of the great Khedive. He immediately returned to the village and told his brothers, cousins and fellow villagers about the failure of his mission. He also told them that for his own part he had decided to leave al-Zaqaziq and seek a new life in Palestine. His determination to migrate was evidently shared by many others who were also unhappy with living conditions in Egypt. Within a few days, a group of about 25 families were already on their way eastwards to the Gaza district in Palestine.[4]

The party, travelling with cattle and animals of burden carrying their provisions and effects, arrived after a few days in the environs of Gaza. They rested for a while and investigated the possibilities for a group of their size in the countryside. Probably they were directed by the Ottoman authorities to reside in Hawj, a small village 18 kilometres to the east of Gaza, which had been reinhabited only a few years earlier on the orders of Mustafa Bey, the governor of Gaza. To encourage settlement in the village, he gave its 21,988 dunums to members of the tribes and to farmers who agreed to live in it, free of charge.[5] The Egyptian newcomers moved to Hawj and, as experienced farmers, set themselves to the task of cultivating the fields around their new village. Their agricultural activity was carried out in partnership with tribesmen of the Tayaha and Tarabin who seemed to have already become the established landowners. A good number of families settled well in the new surroundings, but a few did not stay for long and moved to Gaza town and settled there.[6]

The settlement at Hawj did not, however, for most of the newcomers last for more than 15 years. In 1881 Sattam Ibn Fayiz, shaykh mashayikh Bani Sakhr, invited them to come to his estate at Umm al-ʻAmad, south of Amman, and become his crop-sharing partners. Sattam had earlier, probably in 1867,[7] travelled to Egypt, where he had established contact with the Basil clans in Sharqiya province who were considered to be related to his tribe. He must have been convinced that these farmers, coming from the district he had visited, were experienced farmers upon whom he could rely to be productive. He therefore made an offer to their new leader, Shaykh ʻAbd Allah, son of Muhammad Abu Zayd, to move his clan and join as partners in the newly established villages of Bani Sakhr. Sattam's argument was apparently convincing when he told them that they would fare better with him than with the landlords of Hawj. The fact that they were countrymen of those with whom he claimed kinship in Egypt, he said, brought them closer together. Furthermore, his offer contained an element of special recognition for Shaykh ʻAbd Allah personally. He was to have the right of running three faddans, free of any share payable to Sattam, which meant an average annual saving of between 15 and 20 tons of wheat. His brothers and cousins were also promised good and fair treatment. They were to be charged one-third of the crop only, without any extra amounts for the ʻushr or khawa. In view of these good conditions and the development of a personal relationship between Shaykh Sattam and Shaykh ʻAbd Allah, it is not surprising that ten families agreed to migrate. Shortly before their arrival in Umm al-ʻAmad, other families followed suit and moved to Transjordan. By the 1890s Egyptian farmers were already farming as well in al-Tunayb with al-Junaybat clan, in al-Lubban with al-Bakhit clan, in al-Yaduda with the Abujabers and in Jalul with the Zabn clan.[8] A few families of Egyptian origin also became partners of Sattam in Zizia, the most eastern of Bani Sakhr's agricultural settlements.[9]

It has not been possible to ascertain the number of families who took residence in Umm al-ʻAmad and its neighbouring villages, even though the move took place only just over a hundred years ago and grandsons of the early settlers are still living and can tell the stories they heard. However, there seems to be a general consensus that around 15 men were involved in farming at Umm al-'Amad[10] and smaller numbers in each of the other villages.[11] It also seems to be agreed that the partnership with Sattam, and with his sons after his death in 1890, continued for 12 years.

The pleasant relationship that prevailed between the Egyptians and their landlord-partners encouraged more Egyptians to come and join the fallahin already farming in al-Balqa' district. Members of the Egyptian clans were generally hard working and careful and, under the leadership of Shaykh 'Abd Allah, maintained close contact with each other. Intermarriage between the different families strengthened in them the feeling of being one group that had a common interest.

With time, the leading group at Umm al-'Amad developed a desire to become the cultivators of their own land and not merely crop-sharing partners. They also seem to have had the financial means for this. Sattam and his fellow shaykhs of the Bani Sakhr were not ready to sell land that they had recently acquired and this attitude forced the Egyptians to look for land elsewhere. People at Sahab related that towards the start of 1893, a delegation composed of Shaykh 'Abd Allah Abu Zayd as leader, Mahmud Hasan Abu Zayd and Ibrahim al-Taharwa proceeded to plead with the Ottoman authorities for the purpose of obtaining land. One story was that the delegation, after meeting with no success in Damascus, continued all the way to Istanbul where they were received by the sultan who gave them an audience. Having already acquired the village of Zizia for himself as an imperial domain or *jiftlik*,[12] he liked to see the area around his new lands settled by farmers and was therefore sympathetic to their plea and granted them Sahab with exemption from the arrears of taxes which had accrued.[13] Another story stated that the delegation made its appeal to the wali in Damascus, who agreed to the request and granted the land[14] with the condition that the Egyptians pay the arrears of taxes of 40 Ottoman gold pounds. The two stories are in agreement that some high Ottoman authority agreed to the request made by the Egyptian delegation and granted them the right to

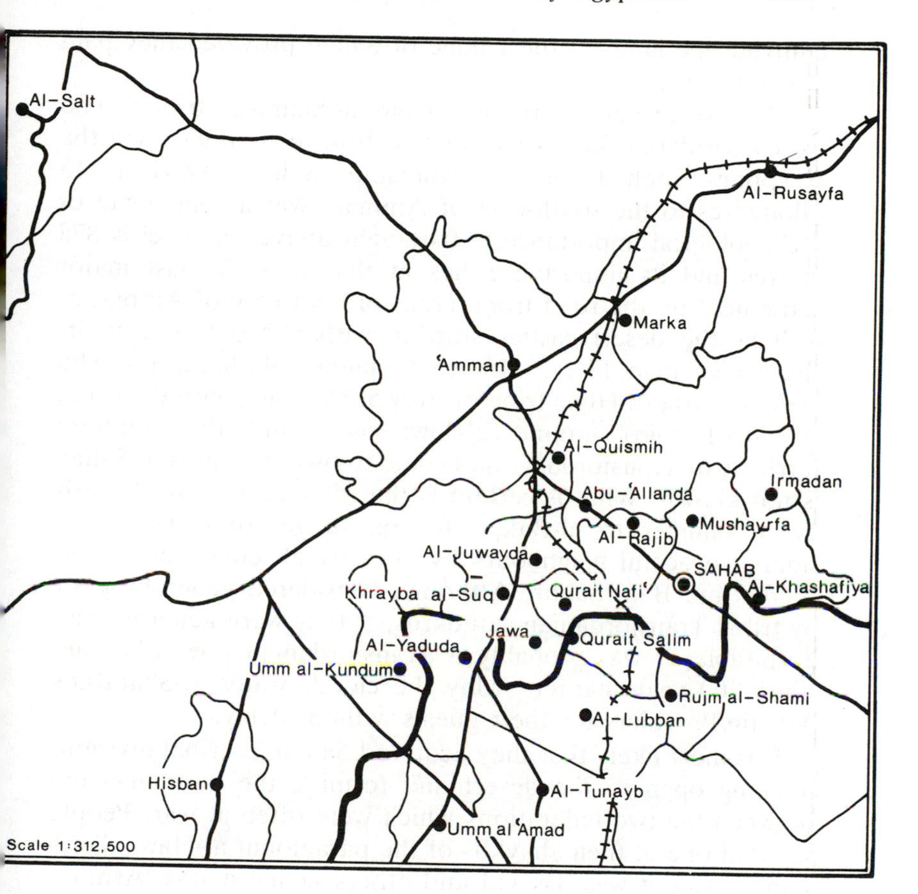

Map 13.1 Situation of Sahab

Note: The dark line is boundary between the Bani Sakhr and al-Balqa' tribal confederation.

cultivate the lands in the village of Sahab provided they paid the ‘ushr every year.

The tax arrears were due from al-Shararat tribe[15] who were considered the owners of Sahab prior to 1893 when the Egyptians applied for its ownership. Sahab, which is 13 kilometres to the south-east of Amman, was a ruined site of archaeological importance.[16] Its height above sea-level is 873 metres and its importance lies in that it is the last major settlement on the road from Transjordan to Saudi Arabia, as well as the desert castles starting with al-Muwaqqar to its south (see map 13.1).[17] In what manner al-Shararat – who were a northern Hijaz tribe of over 5,000 tents, with their dira (tribal domain) around al-Jawf oasis and the southern territory of Transjordan – had come to own the lands of Sahab is not known. What is certain is that they moved northwards every summer to al-Balqa’ for pasturage; they were considered peaceful neighbours by both tribal federations of al-Balqa’ and Bani Sakhr. Although considered generally poor by tribal Transjordanian standards, [18] they were generous and hospitable; it was probably a Transjordanian poet who said some 200 years earlier: ‘May the clouds water al-Shararat’s domain/Providers for their guests without delay.’[19]

It is most likely that they acquired Sahab by tribal division, it being open to the desert and forming the no-man’s-land between the two federations which were often at war. People say that one of their shaykhs of the paramount al-Hawi clan – some saying it was ‘Issayid and others saying it was ‘Ashiq – was known to have cultivated fields in Sahab. The physical work was carried out by oasis dwellers from al-Jawf, possibly vassals of al-Hawi, who employed camels picked from al-Shararat’s herds. This agricultural activity could not have continued for long as the Ottoman authorities were eager to see these fields worked regularly by a settled population. They therefore sent for the shaykh of al-Shararat when the application from the Egyptians was received, and asked him to settle the tax arrears owing on the land. Forty gold pounds was no small amount and al-Hawi naturally told them that he did not have it. The Egyptian delegation expressed their willingness to pay the amount on condition that the shaykh forfeit his tribe’s claim to Sahab. This he willingly did and the authorities allocated the use of the land, a transaction called *tafwid*, to the name of al-Zuyud and al-Taharwa clans.

Another story about Sahab was told in al-Salt, the administrative centre of al-Balqa' during the second half of the nineteenth century.[20] This revealed that, to avoid friction between their members, the tribes of the area agreed to arrange a tribal division of the land between al-Zarqa' and Zizia, a distance of 50 kilometres. A date was set for the shaykhs of the tribes of the Bani Sakhr and al-Balqa' federations and al-Shararat to meet and settle the division. On that day the shaykh of al-Shararat fell asleep suddenly and did not keep the appointment, whereupon the other shaykhs present went ahead with their work and allotted to the absent shaykh's tribe a small area with Sahab as its base. The different clans of the Bani Sakhr were allotted the front south of Sahab to Zizia, while the clans of the Da'ja and al-Balqawiya were allotted the front north of Sahab to al-Zarqa'. Both fronts extended deep into the badiya, and until the 1940s these stretches formed the pasturelands for the herds and flocks of the tribes owning them. Sahab, on the other hand, did not have any front opening on the badiya.

Although some of the details do not stand the test of historical analysis, the stories do confirm that Sahab was a khirba or village and that it was actually taken over by the Egyptians from the Ottoman authorities in 1894. The oldest written report about this settlement operation, and probably the only one until now, is that related by Peake Pasha, who collected the information from first and second generation settlers in the late 1920s.[21] Referring to the Egyptians he wrote:

> It is said that they are descendants of those who stayed behind after the withdrawal of Ibrahim Pasha's army from Palestine. Previously they resided in Gaza and 40 years ago a group migrated to Transjordan and started to cultivate the lands near Madaba. When Husayn Hilmi Pasha was mutasarrif of al-Karak, that is after the Turkish administration was established in the country in 1892,[22] a delegation of them went to Damascus and asked the wali 'Uthman Pasha to give them land which they could own and live on. He granted them the lands on which they later built the two villages Zizia and Sahab.[23]

However, it must be clearly stated that Zizia, which in 1894 was owned by Sultan 'Abd al-Hamid, could not have been given away by any wali, but only by the sultan himself. The Egyptian farmers who lived in Zizia did not therefore have the status of owners but rather that of fallahin in partnership with the sultan's property administration. Later, when the village became wholly the property of Shaykh Mithqal al-Fayiz, the Egyptians at Zizia tried to buy land elsewhere.

Once the legal aspects were settled with the authorities, it became necessary to distribute the 13,643 dunums of land in Sahab. This difficult task was supervised by Shaykh 'Abd Allah Abu Zayd, headman of al-Zuyud and shaykh and spokesman for the whole of Sahab until his death in 1924. One story explained why the people in the village considered its area to be only 12,000 dunums. After coming to terms with the authorities and al-Shararat, the new settlers discovered that clans of al-Da'ja laid claim to fields on the northern side and al-Balqawiya tribes to fields on the western side. Since they were pressed for time, they quickly arrived at a settlement. Areas of about 1,000 dunums were allotted to each tribe in the contested locations.[24] By such firmness and what appeared to be a spirit of neighbourly co-operation, the Egyptians managed to prevent any further claims especially on their southern borders.

Shaykh 'Abd Allah and the leaders among the new settlers were fully aware of the weakness inherent in their situation due to the location of Sahab. They were hemmed in between the two strong tribal federations who had continuous fights about boundaries, pasture rights, and tribal affairs. They strongly believed therefore that their best plan was to remain on the best possible terms with both sides. Since they were farmers, they allowed it to be known that they were part of al-Balqa' federation which included all the settled tribes. However, they maintained especially friendly relations with the Khurshan, shaykhs of the Bani Sakhr clans on their southern and eastern borders. They knew, though, that their numbers and the unity between their clans were the best guarantees for their security. Shaykh 'Abd Allah therefore continued to invite Egyptians as well as others to join his group, even after the ownership issue was settled with the authorities. It is said that on the return of the delegation from its successful mission, al-Maharma clan invited the Egyptians who were still living in Umm al-'Amad to have lunch with them in Jalul, where they were farming the lands of the Zabn clan. Being themselves of Egyptian origin, they felt that their compatriots' success was an occasion for celebration. When the guests arrived at the edge of the encampment, the womenfolk of al-Maharma started to welcome them with *al-zagharit*, or shrill trills of joy, as is customary in the Transjordanian countryside. The guests were deeply touched and spontaneously Shaykh 'Abd Allah announced that the new owners of Sahab would present al-Maharma with a quarter of its lands if they were ready to come and live with them.[25] This was a most generous gift but it demonstrated the pressing need that these new settlers had for another clan to join their ranks. Naturally the young men among them were necessary to cultivate the newly acquired fields but, more importantly, they needed men who were willing to fight for the new way of life that they had chosen to follow.

The task of land distribution was carried out by Shaykh 'Abd Allah in what appears now to be a well-studied and practical way. The fields were divided into two lots: the western, which usually gave a better crop on account of the higher rainfall; and the eastern. The whole area was divided into 100 khanas, each of which was about 100 dunums. The average composition was 40 dunums in western fields and 60 in the eastern. Taking into consideration the number of men in each clan, the 100 khanas, each of which was considered a faddan, were divided into four quarters. Shaykh 'Abd Allah's clan and its affiliated families were allotted one-half of the whole village or 50 faddans. The other two smaller clans were allotted one-quarter or 25 faddans each. Once the land distribution arrangements were completed, the different families on the new frontier gave themselves to the task of developing the resources of their village. Caves were cleaned and used together with the crude houses they built. The heaps of stones that are so common in ruined khirab were used in the building operation, as in similar cases of settlement. The wells and cisterns that had stored water for the inhabitants centuries before were cleaned and used again. The neighbours around became envious of this new activity and differences and fights broke out every now and then; but wisely these were contained in a spirit of firmness, dignity, and co-operation.

The importance of Sahab's story of success lies in the fact that it was the result of a collective decision to own land by a group that knew the difficulties it would encounter. Their choice was the most eastern of the villages in those days, and with full knowledge of the possible consequences they agreed to take up the challenge.

In spite of the difficulties encountered by its people, especially with regard to their relations with the members of the two tribal confederations around Sahab, the new settlement grew in size. By 1910 it had over 200 families and nearly 250 men, who besides cultivating the land were ready to defend themselves, their families, and their property against all incursions.[26] To avoid over-taxation they tried to impress on the authorities at al-Salt and Zizia that they were an integral part of the beduin communities around them. Technically speaking they were member of the Balqa' confederation, but at the same time they maintained pleasant relations with some of their neighbours, especially with al-Khurshan and their shaykh, the renowned Haditha.[27] All the time they stood to gain from the wisdom, courage and far-sightedness of their Shaykh 'Abd Allah, who was without doubt the engineer of this large-scale settlement operation.

Economically they fared well by the standards of the time, as a result of their hard work. In addition to the crops they reaped from the land,[28] averaging a net produce of 500 tons of wheat and 500 tons of barley per season, they were active in developing their flocks. It was confirmed that, in the 1920s, the Egyptians at Sahab were paying animal tax for a declared figure of 22,000 sheep and goats.[29] As usual, a significant number of animals were not accounted for, either through the absence of the owner, or because the whole flock was out in pasture; thus the actual number of their flocks may have been even larger. In this field of economic activity, they shared with the people of Madaba a common interest. Both communities also had an interest in trade and still maintain this, as well as their interest in agriculture and animal breeding though to a different degree from at the start of the twentieth century.[30]

Appendices

Appendix 1
Annual Rainfall Data

Table A Annual rainfall statistical analysis
Independent station X: Jerusalem
Dependent Station Y: Irbid
Linear correlation Y = 18.9 + 0.8 X
Coefficient of correlation: 0.733

Rainfall year	X Actual rainfall (mm)	Y Actual rainfall (mm)	Y Calculated rainfall (mm)	Deviation percentage
1937/38	612.0	490.0	497.4	1.5
1938/39	629.0	748.0	510.7	–31.7
1939/40	509.0	531.0	416.9	–21.5
1940/41	500.0	536.0	409.8	–23.5
1941/42	742.0	816.0	599.1	–26.6
1942/43	718.0	757.0	580.3	–23.3
1943/44	457.0	352.0	376.2	6.9
1944/45	748.0	580.0	603.7	4.1
1945/46	470.0	332.0	386.4	16.4
1946/47	286.0	192.0	242.5	26.3
1947/48	529.0	276.0	432.5	56.7
1948/49	757.0	609.0	610.8	0.3
1949/50	494.0	508.0	405.1	–20.2
1950/51	250.0	312.0	214.3	–31.3
1951/52	673.0	685.0	545.1	–20.4
1952/53	496.0	654.0	406.7	–37.8
1953/54	544.0	376.0	444.2	18.1
1954/55	333.0	193.0	279.2	44.7
1955/56	562.0	415.0	458.3	10.4
1956/57	648.0	344.0	525.6	52.8
1957/58	348.0	322.0	291.0	–9.6
1958/59	420.0	209.0	347.3	66.2

Independent station X: Jerusalem
Dependent Station Y: Irbid
Linear correlation Y = 18.9 + 0.8 X
Coefficient of correlation: 0.733

Rainfall year	X Actual rainfall (mm)	Y Actual rainfall (mm)	Y Calculated rainfall (mm)	Deviation percentage
1959/60	210.0	172.0	183.1	6.4
1960/61	474.0	338.0	389.5	15.2
1961/62	414.0	344.0	342.6	–0.4
1962/63	225.0	248.0	194.8	–21.5
1963/64	687.0	411.0	556.0	35.3
1964/65	628.0	362.0	509.9	40.9
1965/66	335.0	294.0	280.8	–4.5
1966/67	705.0	517.0	570.1	10.3
1967/68	596.0	407.0	484.9	19.1
1968/69	522.0	475.0	427.0	–10.1
1969/70	466.0	409.0	383.2	–6.3
1970/71	531.0	498.0	434.1	–12.8
1971/72	545.0	389.0	445.0	14.4
1972/73	425.0	253.0	351.2	38.8
1973/74	818.0	518.0	658.5	27.1
1974/75	453.0	295.0	373.1	26.5
Mean	520.0	425.4	425.4	22.1
Standard deviation	152.7	162.8	119.4	
Relative dispersion	29.4%	38.3%	28.1%	
Variance	23312.9	26513.2	14253.6	

Covariance of X and Y: 18228.9
Standard error of estimate of Y on X: 110.7
Relative error of estimate of Y on X: 26.0%

Independent station X: Jerusalem
Dependent Station Y: ‘Ajlun
Linear correlation Y = 122.7 + 1.0 X
Coefficient of correlation: 0.791

Rainfall year	X Actual rainfall (mm)	Y Actual rainfall (mm)	Y Calculated rainfall (mm)	Deviation percentage
1937/38	612.0	884.0	746.0	–15.6
1938/39	629.0	750.0	763.3	1.8
1939/40	509.0	742.0	641.1	–13.6

Rainfall year	X Actual rainfall (mm)	Y Actual rainfall (mm)	Y Calculated rainfall (mm)	Deviation percentage
1940/41	500.0	650.0	631.9	–2.8
1941/42	742.0	739.0	878.3	18.9
1942/43	718.0	1031.0	853.9	–17.2
1943/44	457.0	479.0	588.1	22.8
1944/45	748.0	881.0	884.5	0.4
1945/46	470.0	575.0	601.3	4.6
1946/47	286.0	320.0	414.0	29.4
1947/48	529.0	637.0	661.4	3.8
1948/49	757.0	928.0	893.6	–3.7
1949/50	494.0	806.0	625.8	–22.4
1950/51	250.0	308.0	377.3	22.5
1951/52	673.0	904.0	808.1	–10.6
1952/53	496.0	638.0	627.8	–1.6
1953/54	544.0	763.0	676.7	–11.3
1954/55	333.0	384.0	461.8	20.3
1955/56	562.0	724.0	695.0	–4.0
1956/57	648.0	705.0	782.6	11.0
1957/58	348.0	578.0	477.1	–17.5
1958/59	420.0	428.0	550.4	28.6
1959/60	210.0	293.0	336.6	14.9
1960/61	474.0	508.0	605.4	19.2
1961/62	414.0	625.0	544.3	–12.9
1962/63	225.0	565.0	351.8	–37.7
1963/64	687.0	763.0	822.3	7.8
1964/65	628.0	754.0	762.2	1.1
1965/66	335.0	487.0	463.9	–4.8
1966/67	705.0	1024.0	840.7	–17.9
1967/68	596.0	589.0	729.7	23.9
1968/69	522.0	925.0	654.3	–29.3
1969/70	466.0	641.0	597.3	–6.8
1970/71	531.0	710.0	663.5	–6.6
1971/72	545.0	606.0	677.7	11.8
1972/73	425.0	292.0	555.5	90.2
1973/74	818.0	722.0	955.7	32.4
1974/75	453.0	427.0	584.0	36.8
Mean	520.0	652.2	652.2	16.8
Standard deviation	152.7	196.5	155.5	
Relative dispersion	29.4%	30.1%	23.8%	
Variance	23312.9	38596.8	24176.8	

Covariance of X and Y: 23741.0
Standard error of estimate of Y on X: 120.1
Relative error of estimate of Y on X: 18.4%

Independent station X: Jerusalem
Dependent Station Y: Amman
Linear correlation $Y = -12.0 + 0.6\ X$
Coefficient of correlation: 0.930

Rainfall year	X Actual rainfall (mm)	Y Actual rainfall (mm)	Y Calculated rainfall (mm)	Deviation percentage
1937/38	612.0	341.0	335.3	–1.7
1938/39	629.0	343.0	345.0	0.6
1939/40	509.0	286.0	276.9	–3.2
1940/41	500.0	214.0	271.8	27.0
1941/42	742.0	402.0	409.1	1.8
1942/43	718.0	380.0	395.5	4.1
1943/44	457.0	222.0	247.4	11.4
1944/45	748.0	480.0	412.5	–14.1
1945/46	470.0	287.0	254.7	–11.2
1946/47	286.0	137.0	150.3	9.7
1947/48	529.0	295.0	288.2	–2.3
1948/49	757.0	428.0	417.6	–2.4
1949/50	494.0	302.0	268.4	–11.1
1950/51	250.0	136.0	129.9	–4.5
1951/52	673.0	363.0	369.9	1.9
1952/53	496.0	306.0	269.5	–11.9
1953/54	544.0	290.0	296.7	2.3
1954/55	333.0	154.0	177.0	14.9
1955/56	562.0	331.0	307.0	–7.3
1956/57	648.0	358.0	355.8	–0.6
1957/58	348.0	220.0	185.5	–15.7
1958/59	420.0	206.0	226.4	9.9
1959/60	210.0	104.0	107.2	3.1
1960/61	474.0	249.0	257.0	3.2
1961/62	414.0	247.0	223.0	–9.7
1962/63	225.0	141.0	115.7	–17.9
1963/64	687.0	431.0	377.9	10.8
1964/65	628.0	272.0	344.4	26.6
1965/66	335.0	218.0	178.1	–18.3
1966/67	705.0	462.0	388.1	–16.0
1967/68	596.0	254.0	326.2	28.4
1968/69	522.0	323.0	284.3	–12.0
1969/70	466.0	176.0	252.5	43.5
1970/71	531.0	294.0	289.4	–1.6
1971/72	545.0	313.0	297.3	–5.0
1972/73	425.0	195.0	229.2	17.5
1973/74	818.0	448.0	452.2	0.9
1974/75	453.0	240.0	245.1	2.1

Rainfall year	X Actual rainfall (mm)	Y Actual rainfall (mm)	Y Calculated rainfall (mm)	Deviation percentage
Mean	520.0	283.1	283.1	10.2
Standard deviation	152.7	93.2	86.6	
Relative dispersion	29.4%	32.9%	30.6%	
Variance	23312.9	8681.2	7506.9	

Covariance of X and Y: 13229.1
Standard error of estimate of Y on X: 34.3
Relative error of estimate of Y on X: 12.1%

Independent station X: Jerusalem
Dependent Station Y: Salt
Linear correlation Y = 2.9 + 1.2 X
Coefficient of correlation: 0.856

Rainfall year	X Actual rainfall (mm)	Y Actual rainfall (mm)	Y Calculated rainfall (mm)	Deviation percentage
1937/38	612.0	997.0	747.5	–25.0
1938/39	629.0	852.0	768.2	–9.8
1939/40	509.0	783.0	622.2	–20.5
1940/41	500.0	644.0	611.2	–5.1
1941/42	742.0	894.0	905.6	1.3
1942/43	718.0	1005.0	876.4	–12.8
1943/44	457.0	508.0	558.9	10.0
1944/45	748.0	1038.0	912.9	–12.0
1945/46	470.0	684.0	574.7	–16.0
1946/47	286.0	366.0	350.9	–4.1
1947/48	529.0	659.0	646.5	–1.9
1948/49	757.0	939.0	923.9	–1.6
1949/50	494.0	772.0	603.9	–21.8
1950/51	250.0	326.0	307.1	–5.8
1951/52	673.0	926.0	821.7	–11.31
1952/53	496.0	647.0	606.4	–6.3
1953/54	544.0	682.0	664.8	–2.5
1954/55	333.0	287.0	408.1	42.2
1955/56	562.0	643.0	686.7	6.8
1956/57	648.0	651.0	791.3	21.5
1957/58	348.0	355.0	426.3	20.1
1958/59	420.0	466.0	513.9	10.3
1959/60	210.0	302.0	258.4	–14.4

Independent station X: Jerusalem
Dependent Station Y: Salt
Linear correlation Y = 2.9 + 1.2 X
Coefficient of correlation: 0.856

Rainfall year	X Actual rainfall (mm)	Y Actual rainfall (mm)	Y Calculated rainfall (mm)	Deviation percentage
1960/61	474.0	654.0	579.6	11.4
1961/62	414.0	433.0	506.6	17.0
1962/63	225.0	339.0	276.7	–18.4
1963/64	687.0	756.0	838.7	10.9
1964/65	628.0	633.0	766.9	21.2
1965/66	335.0	427.0	410.5	–3.9
1966/67	705.0	918.0	860.6	–6.3
1967/68	596.0	412.0	728.0	76.7
1968/69	522.0	721.0	638.0	–11.5
1969/70	466.0	423.0	569.9	34.7
1970/71	531.0	718.0	648.9	–9.6
1971/72	545.0	640.0	666.0	4.1
1972/73	425.0	350.0	520.0	48.6
1973/74	818.0	796.0	998.1	25.4
1974/75	453.0	504.0	554.0	9.9
Mean	520.0	635.5	635.5	15.6
Standard deviation	152.7	217.1	185.8	
Relative dispersion	29.4%	34.2%	29.2%	
Variance	23312.9	47131.8	34504.7	

Covariance of X and Y: 28362.0
Standard error of estimate of Y on X: 112.4
Relative error of estimate of Y on X: 17.7%

Independent station X: Jerusalem
Dependent Station Y: Madaba
Linear correlation Y = 12.8 + 0.7 X
Coefficient of correlation: 0.747

Rainfall year	X Actual rainfall (mm)	Y Actual rainfall (mm)	Y Calculated rainfall (mm)	Deviation percentage
1937/38	612.0	397.0	411.7	3.7
1938/39	629.0	397.0	422.8	6.5
1939/40	509.0	448.0	344.6	–23.1

Rainfall year	X Actual rainfall (mm)	Y Actual rainfall (mm)	Y Calculated rainfall (mm)	Deviation percentage
1940/41	500.0	257.0	338.7	31.8
1941/42	742.0	550.0	496.4	–9.7
1942/43	718.0	452.0	480.8	6.4
1943/44	457.0	295.0	310.7	5.3
1944/45	748.0	616.0	500.3	–18.8
1945/46	470.0	350.0	319.1	–8.8
1946/47	286.0	170.0	199.2	17.2
1947/48	529.0	350.0	357.6	2.2
1948/49	757.0	500.0	506.2	1.2
1949/50	494.0	640.0	334.8	–47.7
1950/51	250.0	204.0	175.7	–13.9
1951/52	673.0	472.0	451.5	–4.4
1952/53	496.0	309.0	336.1	8.8
1953/54	544.0	415.0	367.4	–11.5
1954/55	333.0	221.0	229.8	4.0
1955/56	562.0	417.0	379.1	–9.1
1956/57	648.0	518.0	435.2	–16.0
1957/58	348.0	141.0	239.6	69.9
1958/59	420.0	158.0	286.5	81.4
1959/60	210.0	268.0	149.7	–44.2
1960/61	474.0	324.0	321.7	–0.7
1961/62	414.0	336.0	282.6	–15.9
1962/63	225.0	136.0	159.4	17.2
1963/64	687.0	513.0	460.6	–10.2
1964/65	628.0	452.0	422.1	–6.6
1965/66	335.0	226.0	231.1	2.3
1966/67	705.0	447.0	472.3	5.7
1967/68	596.0	351.0	401.3	14.3
1968/69	522.0	475.0	353.0	–25.7
1969/70	466.0	225.0	316.5	40.7
1970/71	531.0	311.0	358.9	–15.4
1971/72	545.0	352.0	368.0	4.6
1972/73	425.0	143.0	289.8	102.7
1973/74	818.0	339.0	546.0	61.1
1974/75	453.0	190.0	308.1	62.1
Mean	520.0	351.7	351.7	21.9
Standard deviation	152.7	133.2	99.5	
Relative dispersion	29.4%	37.9%	28.3%	
Variance	23312.9	17743.3	9905.0	

Covariance of X and Y: 15195.9
Standard error of estimate of Y on X: 88.5
Relative error of estimate of Y on X: 25.2%

Independent station X: Jerusalem
Dependent Station Y: Karak
Linear correlation Y = –12.8 + 0.7 X
Coefficient of correlation: 0.811

Rainfall year	X Actual rainfall (mm)	Y Actual rainfall (mm)	Y Calculated rainfall (mm)	Deviation percentage
1937/38	612.0	408.0	427.1	4.7
1938/39	629.0	424.0	439.3	3.6
1939/40	509.0	522.0	353.0	–32.4
1940/41	500.0	398.0	346.6	–12.9
1941/42	742.0	541.0	520.5	–3.8
1942/43	718.0	447.0	503.2	12.6
1943/44	457.0	403.0	315.7	–21.7
1944/45	748.0	524.0	524.8	0.2
1945/46	470.0	358.0	325.0	–9.2
1946/47	286.0	123.0	192.8	56.7
1947/48	529.0	293.0	367.4	25.4
1948/49	757.0	464.0	531.3	14.5
1949/50	494.0	280.0	342.3	22.2
1950/51	250.0	215.0	166.9	–22.4
1951/52	673.0	500.0	470.9	–5.8
1952/53	496.0	215.0	343.7	59.9
1953/54	544.0	360.0	378.2	5.1
1954/55	333.0	230.0	226.6	–1.5
1955/56	562.0	452.0	391.1	–13.5
1956/57	648.0	414.0	452.9	9.4
1957/58	348.0	156.0	237.3	52.1
1958/59	420.0	282.0	289.1	2.5
1959/60	210.0	145.0	138.2	–4.7
1960/61	474.0	273.0	327.9	20.1
1961/62	414.0	210.0	284.8	35.6
1962/63	225.0	102.0	148.9	46.0
1963/64	687.0	496.0	481.6	–3.0
1964/65	628.0	661.0	438.6	–33.7
1965/66	335.0	321.0	228.0	–29.0
1966/67	705.0	438.0	493.9	12.8
1967/68	596.0	387.0	415.6	7.4
1968/69	522.0	359.0	362.4	0.9
1969/70	466.0	291.0	322.1	10.7
1970/71	531.0	357.0	368.8	3.3
1971/72	545.0	606.0	378.9	–37.5
1972/73	425.0	187.0	292.7	56.5
1973/74	818.0	509.0	575.1	13.0
1974/75	453.0	364.0	312.8	–14.1

Rainfall year	X Actual rainfall (mm)	Y Actual rainfall (mm)	Y Calculated rainfall (mm)	Deviation percentage
Mean	520.0	360.9	360.9	19.0
Standard deviation	152.7	135.4	109.7	
Relative dispersion	29.4%	37.5%	30.4%	
Variance	23312.9	18325.3	12039.8	

Covariance of X and Y: 16753.6
Standard error of estimate of Y on X: 79.3
Relative error of estimate of Y on X: 22.0%

Independent station X: Jerusalem
Dependent Station Y: Tafila
Linear correlation Y = –7.1 + 0.6 X
Coefficient of correlation: 0.608

Rainfall year	X Actual rainfall (mm)	Y Actual rainfall (mm)	Y Calculated rainfall (mm)	Deviation percentage
1937/38	612.0	354.0	336.5	–5.0
1938/39	629.0	325.0	346.0	6.5
1939/40	509.0	327.0	278.7	–14.8
1940/41	500.0	209.0	273.6	30.9
1941/42	742.0	306.0	409.5	33.8
1942/43	718.0	323.0	396.0	22.6
1943/44	457.0	346.0	249.5	–27.9
1944/45	748.0	464.0	412.8	–11.0
1945/46	470.0	324.0	256.8	–20.8
1946/47	286.0	131.0	153.5	17.2
1947/48	529.0	267.0	289.9	8.6
1948/49	757.0	268.0	417.9	55.9
1949/50	494.0	380.0	270.2	–28.9
1950/51	250.0	269.0	133.3	–50.5
1951/52	673.0	304.0	370.7	21.9
1952/53	496.0	106.0	271.4	156.0
1953/54	544.0	301.0	298.3	–0.9
1954/55	333.0	150.0	179.9	19.9
1955/56	562.0	215.0	308.4	43.4
1956/57	648.0	363.0	356.7	–1.7
1957/58	348.0	89.0	188.3	111.5
1958/59	420.0	384.0	228.7	–40.4
1959/60	210.0	87.0	110.8	27.4
1960/61	474.0	204.0	259.0	27.0
1961/62	414.0	136.0	225.3	65.7

Independent station X: Jerusalem
Dependent Station Y: Tafila
Linear correlation Y = –7.1 + 0.6 X
Coefficient of correlation: 0.608

Rainfall year	X Actual rainfall (mm)	Y Actual rainfall (mm)	Y Calculated rainfall (mm)	Deviation percentage
1962/63	225.0	84.0	119.2	41.9
1963/64	687.0	617.0	378.6	–38.6
1964/65	628.0	751.0	345.5	–54.0
1965/66	335.0	167.0	181.0	8.4
1966/67	705.0	343.0	388.7	13.3
1967/68	596.0	282.0	327.5	16.1
1968/69	522.0	295.0	286.0	–3.1
1969/70	466.0	132.0	254.5	92.8
1970/71	531.0	257.0	291.0	13.2
1971/72	545.0	439.0	298.9	–31.9
1972/73	425.0	82.0	231.5	182.3
1973/74	818.0	403.0	452.1	12.2
1974/75	453.0	339.0	247.2	–27.1
Mean	520.0	284.8	284.8	36.5
Standard deviation	152.7	141.0	85.7	
Relative dispersion	29.4%	49.5%	30.1%	
Variance	23312.9	19891.2	7346.6	

Covariance of X and Y: 13087.0
Standard error of estimate of Y on X: 112.0
Relative error of estimate of Y on X: 39.3%

Independent station X: Jerusalem
Dependent Station Y: Shaubak
Linear correlation Y = –2.5 + 0.6 X
Coefficient of correlation- 0.687

Rainfall year	X Actual rainfall (mm)	Y Actual rainfall (mm)	Y Calculated rainfall (mm)	Deviation percentage
1937/38	612.0	443.0	387.5	–12.5
1938/39	629.0	362.0	398.4	10.0
1939/40	509.0	325.0	321.9	–1.0
1940/41	500.0	346.0	316.1	–8.6
1941/42	742.0	696.0	470.4	–32.4
1942/43	718.0	702.0	455.1	–35.2
1943/44	457.0	435.0	288.7	–33.6
1944/45	748.0	529.0	474.2	–10.4
1945/46	470.0	298.0	297.0	–0.3
1946/47	286.0	232.0	179.8	–22.5
1947/48	529.0	326.0	334.6	2.6
1948/49	757.0	336.0	479.9	42.8
1949/50	494.0	392.0	312.3	–20.3
1950/51	250.0	278.0	156.8	–43.6
1951/52	673.0	361.0	426.4	18.1
1952/53	496.0	138.0	313.6	127.2
1953/54	544.0	378.0	344.2	–8.9
1954/55	333.0	234.0	209.7	–10.4
1955/56	562.0	274.0	355.7	29.8
1956/57	648.0	318.0	410.5	29.1
1957/58	348.0	128.0	219.3	71.3
1958/59	420.0	286.0	265.2	–7.3
1959/60	210.0	175.0	131.3	–25.0
1960/61	474.0	251.0	299.6	19.4
1961/62	414.0	236.0	261.3	10.7
1962/63	225.0	115.0	140.9	22.5
1963/64	687.0	558.0	435.3	–22.0
1964/65	628.0	521.0	397.7	–23.7
1965/66	335.0	269.0	211.0	–21.6
1966/67	705.0	412.0	446.8	8.4
1967/68	596.0	330.0	377.3	14.3
1968/69	522.0	357.0	330.2	–7.5
1969/70	466.0	161.0	294.5	82.9
1970/71	531.0	206.0	335.9	63.1
1971/72	545.0	429.0	344.8	–19.6
1972/73	425.0	75.0	268.3	257.8
1973/74	818.0	345.0	518.8	50.4
1974/75	453.0	240.0	286.2	19.2

Independent station X: Jerusalem
Dependent Station Y: Shaubak
Linear correlation Y = –2.5 + 0.6 X
Coefficient of correlation- 0.687

Rainfall year	X Actual rainfall (mm)	Y Actual rainfall (mm)	Y Calculated rainfall (mm)	Deviation percentage
Mean	520.0	328.9	328.9	32.8
Standard deviation	152.7	141.6	97.3	
Relative dispersion	29.4%	43.0%	29.6%	
Variance	23312.9	20043.1	9468.9	

Covariance of X and Y: 14857.6
Standard error of estimate of Y on X: 102.8
Relative error of estimate of Y on X: 31.3%

Table B Annual rainfall statistical summary: Jerusalem and Transjorden (1937–75)

	Jerusalem	Irbid	ʻAjlun	Salt	Amman	Madaba	Karak	Tafila	Shaubak
Latitude (degrees east)	31:47	32:33	32:20	32:02	31:58	31:43	31:11	30:50	30:32
Longitude (degrees north)	35:11	35:51	35:45	35:44	35:56	35:48	35:42	35:36	35:31
Altitude (metres)	740	585	760	796	790	785	920	1000	992
Mean annual rainfall (mm)	520	425	652	636	283	352	361	285	329
Standard deviation of annual rainfall distribution (mm)	153	163	196	217	93	133	135	141	142
Relative deviation of annual rainfall distribution	29%	38%	30%	34%	33%	38%	38%	50%	43%
Correlation coefficient with Jerusalem rainfall	1.00	0.73	0.79	0.86	0.93	0.75	0.81	0.61	0.69
Standard error of estimate from Jerusalem rainfall data (mm)	0	111	10	112	34	89	79	112	103
Relative error of estimate from Jerusalem rainfall data	0%	26%	18%	18%	12%	25%	22%	39%	31%

Table C Annual rainfall data for Jerusalem and Transjordan

Water year	Jerusalem	Amman	Irbid	‘Ajlun	Salt	Madaba	Karak	Tafila	Shaubak
1846/1847	477	258*	391*	608*	583*	323*	330*	260*	301*
1847/1848	412	221*	341*	542*	504*	281*	283*	224*	259*
1848/1849	490	265*	402*	621*	599*	332*	339*	267*	309*
1849/1850	560	305*	456*	692*	684*	377*	389*	307*	354*
1850/1851	687	377*	556*	822*	839*	460*	481*	378*	435*
1851/1852	526	286*	430*	658*	643*	355*	365*	288*	332*
1852/1853	356	189*	297*	485*	436*	244*	243*	192*	224*
1853/1854	700	384*	566*	835*	854*	469*	490*	385*	443*
1854/1855	542	295*	442*	674*	662*	366*	376*	296*	342*
1855/1856	640	350*	519*	774*	781*	430*	447*	351*	405*
1856/1857	794	438*	639*	931*	969*	530*	558*	438*	503*
1957/1858	636	348*	516*	770*	776*	427*	444*	349*	402*
1858/1859	522	284*	427*	654*	638*	353*	362*	285*	329*
1859/1860	560	305*	456*	692*	684*	377*	389*	307*	354*
1860/1861	557	303*	454*	689*	680*	375*	387*	305*	352*
1861/1862	623	341*	506*	756*	761*	418*	435*	342*	394*
1862/1863	588	321*	478*	721*	718*	396*	410*	322*	372*
1863/1864	523	284*	427*	655*	639*	353*	363*	286*	330*
1864/1865	419	225*	346*	549*	512*	285*	288*	227*	264*
1865/1866	486	263*	398*	617*	594*	329*	336*	265*	307*

1866/1867	643	352*	521*	777*	785*	432*	449*	353*	407*
1867/1868	689	378*	557*	824*	841*	462*	482*	379*	436*
1868/1869	633	346*	513*	767*	773*	425*	442*	348*	400*
1869/1870	318	168*	267*	446*	389*	220*	215*	171*	200*
1870/1871	486	263*	398*	617*	594*	329*	336*	265*	307*
1871/1872	469	253*	385*	600*	573*	318*	324*	256*	296*
1872/1873	481	260*	395*	612*	588*	326*	333*	262*	303*
1873/1874	1004	557*	803*	1144*	1224*	667*	709*	556*	637*
1874/1875	676	371*	547*	810*	825*	453*	473*	372*	428*
1875/1876	419	225*	346*	549*	512*	285*	288*	227*	264*
1876/1877	353	188*	294*	482*	432*	242*	241*	190*	222*
1877/1878	1090	606*	871*	1232*	1329*	723*	770*	604*	691*
1878/1879	409	219*	338*	539*	500*	279*	281*	222*	258*
1879/1880	598	327*	486*	731*	730*	402*	417*	328*	378*
1880/1881	675	370*	546*	809*	824*	452*	472*	371*	427*
1881/1882	598	327*	486*	731*	730*	402*	417*	328*	378*
1882/1883	584	319*	475*	717*	713*	393*	407*	320*	369*
1883/1884	718	395*	580*	853*	876*	480*	503*	395*	454*
1884/1885	575	314*	468*	708*	702*	387*	400*	315*	363*
1885/1886	638	349*	517*	772*	779*	428*	445*	350*	403*
1886/1887	674	370*	545*	808*	823*	452*	471*	371*	426*
1887/1888	433	233*	357*	563*	529*	295*	298*	235*	273*
1888/1889	731	402*	590*	866*	892*	489*	512*	403*	463*

(Table continued)

Water year	Jerusalem	Amman	Irbid	ʻAjlun	Salt	Madaba	Karak	Tafila	Shaubak
1889/1890	524	285*	428*	656*	640*	354*	364*	286*	331*
1890/1891	742	408*	599*	878*	905*	496*	520*	409*	470*
1891/1892	646	354*	524*	780*	789*	433*	451*	355*	408*
1892/1893	825	455*	664*	962*	1006*	550*	580*	455*	523*
1893/1894	630	345*	511*	764*	769*	423*	440*	346*	398*
1894/1895	518	281*	423*	650*	633*	350*	359*	283*	327*
1895/1896	770	424*	621*	906*	940*	514*	540*	424*	487*
1896/1897	796	439*	641*	933*	971*	531*	559*	439*	504*
1897/1898	632	346*	513*	766*	772*	424*	441*	347*	400*
1898/1899	525	285*	429*	657*	641*	355*	364*	287*	331*
1899/1900	499	270*	409*	630*	610*	338*	346*	272*	315*
1900/1901	339	180*	283*	467*	415*	233*	230*	183*	213*
1901/1902	509	276*	416*	640*	622*	344*	353*	278*	321*
1902/1903	678	372*	549*	812*	828*	454*	474*	373*	429*
1903/1904	448	242*	369*	578*	548*	304*	309*	244*	282*
1904/1905	794	438*	639*	931*	969*	530*	558*	438*	503*
1905/1906	828	457*	666*	965*	1010*	552*	582*	457*	524*
1906/1907	441	238*	363*	571*	539*	300*	304*	240*	278*
1907/1908	633	346*	513*	767*	773*	425*	442*	348*	400*
1908/1909	611	334*	496*	744*	746*	411*	426*	335*	386*
1909/1910	544	296*	444*	676*	664*	367*	378*	298*	344*
1910/1911	786	433*	633*	922*	959*	525*	552*	433*	498*

1911/1912	572	312*	466*	705*	699*	385*	398*	313*	361*
1912/1913	547	298*	446*	679*	668*	369*	380*	299*	345*
1913/1914	434	234*	358*	564*	531*	295*	299*	236*	273*
1914/1915	451	243*	371*	581*	551*	306*	311*	245*	284*
1915/1916	524	285*	428*	656*	640*	354*	364*	286*	331*
1916/1917	480	260*	394*	611*	587*	325*	332*	262*	303*
1917/1918	492	267*	403*	623*	601*	333*	340*	268*	310*
1918/1919	552	301*	450*	684*	674*	372*	384*	302*	349*
1919/1920	787	434*	634*	923*	960*	525*	553*	434*	498*
1920/1921	567	309*	462*	699*	692*	382*	394*	311*	358*
1921/1922	569	310*	463*	701*	695*	383*	396*	312*	359*
1922/1923	392	210*	325*	521*	480*	268*	269*	212*	247*
1923/1924	496	269*	406*	627*	606*	336*	343*	271*	313*
1924/1925	278	145*	236*	405*	341*	194*	187*	148*	174*
1925/1926	476	257*	391*	607*	582*	323*	329*	259*	300*
1926/1927	499	270*	409*	630*	610*	338*	346*	272*	315*
1927/1928	376	201*	312*	505*	460*	257*	257*	203*	236*
1928/1929	556	303*	453*	688*	679*	375*	387*	304*	351*
1929/1930	440	237*	362*	570*	538*	299*	303*	239*	277*
1930/1931	433	233*	357*	563*	529*	295*	298*	235*	273*
1931/1932	319	168*	268*	447*	391*	220*	216*	171*	200*

(Table continued)

Water year	Jerusalem	Amman	Irbid	'Ajlun	Salt	Madaba	Karak	Tafila	Shaubak
1932/1933	249	129*	213*	376*	305*	175*	166*	132*	156*
1933/1934	375	200*	312*	504*	459*	257*	256*	203*	236*
1934/1935	482	261*	395*	613*	589*	327*	333*	263*	304*
1935/1936	349	185*	291*	477*	427*	240*	238*	188*	219*
1936/1937	639	350*	518*	773*	780*	429*	446*	351*	404*
1937/1938	612	341	490	884	997	397	408	354	443
1938/1939	629	343	748	750	852	397	424	325	362
1939/1940	509	286	531	742	783	448	552	327	325
1940/1941	500	214	536	650	644	257	398	209	346
1941/1942	742	402	816	739	894	550	541	306	696
1942/1943	718	380	757	1031	1005	452	447	323	702
1943/1944	457	222	352	479	508	295	403	346	435
1944/1945	748	480	580	881	1038	616	524	464	529
1945/1946	470	287	332	575	684	350	358	324	298
1946/1947	286	137	192	320	366	170	123	131	232
1947/1948	529	295	276	637	6599	350	293	267	326
1948/1949	757	428	609	928	939	500	464	268	336
1949/1950	494	302	508	806	772	640	280	380	392
1950/1951	250	136	312	308	326	204	215	269	278
1951/1952	673	363	685	904	926	472	500	304	361

1952/1953	496	306	654	638	647	309	215	106	138
1953/1954	544	290	376	763	682	415	360	301	378
1954/1955	333	154	193	384	287	221	230	150	234
1955/1956	562	331	415	724	643	417	452	215	274
1956/1957	648	358	344	705*	651	518	414	363	318
1957/1958	348	220	322	578	355	141	156	89	128
1958/1959	420	206	209	428	466	158	282	384	286
1959/1960	210	104	172	293	302	268	145	87	175
1960/1961	474	249	338	508	654	324	273	204	251
1961/1962	414	247	344	625	433	336	210	136	236
1963/1963	225	141	248	565	339	136	102	84	115
1963/1964	687	341	411	763	756	513	496	617	558
1964/1965	628	272	362	754	633	452	661	751	521
1965/1966	335	218	294	487	427	226	321	167	269
1966/1967	705	462	517	1024	918	447	438	343	412
1967/1968	596	254	407	589	412	351	387	282	330
1968/1969	522	323	475	925	721	475	359	295	357
1969/1970	466	176	409	641	423	225	291	132	161
1970/1971	531	294	498	710	718	311	357	257	206
1971/1972	545	313	389	606	640	352	606	439	429
1972/1973	425	195	253	292	350	143	187	82	75
1973/1974	818	448	518	722	796	339	509	403	345
1974/1975	453	240	295	427	504	190	364	339	240

Note: (*) denotes estimated value.

Appendix 2
Measurements, Weights, Prices and Currency

Measures of land area

qirat a share of property owned in common with one or more partners. Generally the whole was 24 qirats.
faddan the word is applied to both a yoke of oxen and the area cultivated by it during a whole season. Its area varied but in Transjordan was between 8 and 15 hectares.
dunum a specific area which continued to be 1,000 square pics until the 1930s. After that its area was fixed at 1,000 square metres.
pic a measure varying between 18 and 28 inches.

Measures and weights

Containers with which cereals were measured were made from the bark of poplar trees. The most commonly used were:
sa' equal to 6 kg of wheat or 5.5 kg of barley
kayl equal to 12 sa's or 72 kg
mudd equal to 3 sa's or 18 kg

Weights in general use were:
'awqiya about 250 grams or a twelfth of a rutl
rutl about 3 kg or 12 'awqiyas
qintar about 300 kg or 100 rutls

Prices

As Transjordan was greatly underpopulated during the

nineteenth century, commercial activity was at the lowest possible standard and was limited to barter trade. Prices varied from place to place and increased steadily due to inflation and shortages of supply. At the turn of the century, prices for different commodities were:

Unit	*Product*	*Price in piastres*
sa'	wheat	around 6
sa'	barley	around 3
pic	cotton cloth	3–4
pair	shoes	18–24
rutl	raisins	3–4
rutl	onions	around 1
one	lamb or kid	around 24
pair	oxen	750–800
one	camel	300–400

Currency

Ottoman currency was in circulation and the gold pound or lira was the unit equivalent to 100 piastres. A silver coin, the Majidi, was in common use during the nineteenth century and its value varied between 20 and 24 piastres. Foreign gold pounds were also in use: the sterling gold pound was worth about 130 piastres, while the French gold pound was worth about 114 piastres.

Notes

Chapter 1

1. Tristram, *Moab*, p. 4. In 1872, Dr H. B. Tristram headed the expedition organized by the British Association, at its meeting in Edinburgh, for the purpose of undertaking a geographical exploration of the country of Moab.
2. Merrill, *East of Jordan*, p. vi.
3. Schumacher, *Northern Ajlun*, p. 28.
4. Emir 'Abdallah, *Mudhakkirat*, p. 189.
5. This map was published by E. Mavromatis in October 1922 in a paper called 'Preliminary Scheme for the Irrigation of the Valley of the Jordan'.
6. Libbey and Hoskins, *The Jordan Valley and Petra*. p. 32.
7. *Colonial Report No. 20* to the League of Nations on the administration of Palestine and Transjordan for the year 1925, p. 60.
8. Bell, *The Desert and the Sown*; Nelson, ed., *The Desert and the Sown*.
9. Al-Muqaddasi, *Ahsan al-Taqasim*, p. 175.
10. *Colonial Report No. 20* to the League of Nations on the administration of Palestine and Transjordan for the year 1925, p. 62.
11. Merrill, *East of the Jordan*, p. 467.
12. Ionides, *Water Resources*. Wells were still being dug out, as I remember distinctly, during the 1930s. In those days workmen from al-Jawf oasis, who came to al-Balqa' seeking employment, were the most expert in this type of work.
13. It has not been possible to trace such data, even in the salnames (Ottoman yearbooks) of the wilaya of Damascus.
14. The *World Weather Records*, vol. 2, gives figures for Amman between 1951 and 1960, whereas it gives figures for Jerusalem between 1861 and 1960. However, it contains the following reservation concerning Jerusalem precipitation 1846–1914 (p. 184): 'For varying intervals the monthly or daily data from the Old City Station had to be corrected for

faulty interpretation of measurements and for measurements with faulty instruments by comparison with synchronous measurements at other stations or with other instruments at the same station.'

15. *30 Year Average Rainfall*, Technical paper no. 34, part I, p. 1. The Hijaz railway, a major Ottoman and Muslim enterprise, was started for military and political reasons in 1900 and, when it was inaugurated in September 1908, connected Damascus with Madina, a distance of 800 miles.
16. Lawrence, *Seven Pillars*, p. 523.
17. *30 Year Average Rainfall*, Technical Paper no. 34, part I, p. 1; Roseman, 'One hundred years', p. 140.
18. Glaisher, *Meteorological Observations*, p. 19.
19. The document for this series was kindly provided by Mr C. G. Smith in late October 1985.
20. *Rainfall in Jordan*, Technical Paper no. 42, table 4.
21. Blake and Goldschmidt, *Geology*, p. 2.
22. I wish to recognize, with gratitude, the assistance so kindly offered by the computer expert Mr Munir M.Qawar, in Amman.

Chapter 2

1. Farid, *Tarikh*, pp. 193–9.
2. Ma'oz, *Ottoman Reform*, p. 2.
3. Seetzen, Brief Account, p. 35.
4. Buckingham, *Travels*, p. 28.
5. This story was related by a great-aunt, Tamam Abujaber Qub'ayn, in 1936 when she was nearly 86 years old. It was confirmed by my uncle Sa'id Abujaber, in 1964, who had heard it from his grandfather Salih.
6. A descendant of this mason was the Reverend Najib al-Far who worked for over 30 years as Protestant pastor in al-Salt and later during the 1960s in Amman. A branch of the family is still living in Acre and Nazareth, where some of them came to be known by a new name, Habayib.
7. Rafeq, *Province of Damascus*, p. 213. Probably the son of Qa'dan al-Fayiz, who at the head of the Bani Sakhr tribesmen in 1757 carried out the disastrous attack on the *jarda* (relief column) and the pilgrimage caravan resulting in great loss of life and property.
8. Al-'Awra, *Tarikh*, pp. 101, 127.
9. Rafeq, *Province of Damascus*, p. 198.
10. Al-Husari, *Al-bilad al-'arabiya*, p. 59.
11. Burckhardt, *Travels*, p. 383.
12. Irby and Mangles, *Travels*, p. 320.
13. Wallin, *Suwar*, p. 7. Many travellers, while travelling in the East, disguised themselves as people of the country and adopted local names.
14. Ma'oz, *Ottoman Reform*, pp. 19, 150.
15. Ghawanma, *Sharq al-urdunn*, political section, p. 50.
16. Tusun, *Al-jaysh al-misri*, p. 95. The effort to build ships in Egypt

with French technical expertise and French workmen at the arsenal in Alexandria was so great that the first 100-gun battleship was launched on 3 January 1831; ibid., p. 96.

17. Rustum, *Ara'*, p. 154; *idem*, *Al-Mahfuzat*, vol. III, p. 101. Daftar 77 no. 59 dated 8 Dhü al-Hija 1251 H. Dispatch requesting preparation of ships for transporting soldiers and equipment from Alexandria to Syria.

18. Ibid., vol. II, p. 53, quoted Butrus Karami as also saying in July 1832: 'Conditions in this area are stable but beduins are practising aggression and over-exaction in Hawran, Irbid, and 'Ajlun, through pillaging, banditry and highway robbery.'

19. Tusun, *Al-jaysh al-misri*, p. 168, confirms that the nearest garrisons to Transjordan were the 25th Infantry Brigade at Jerusalem (1,755 officers and soldiers) and heavily armed 2nd Cavalry Brigade at Baysan (844 officers and soldiers); Sipano ed., *Mudhakkirat*, vol. 2, p. 152, gives the list in a concise form, probably based on the same sources but without mention of the brigade in Baysan.

20. A strong tribe whose shaykhs Ibn Smayr and al-Tayyar co-operated closely with the Egyptians and later with the Ottomans. Their domain was the Hawran and the areas east of Damascus.

21. Another strong tribe which in 1806, according to Burckhardt, defeated a force of 6,000 soldiers sent against them by the wali of Baghdad. Their shaykhs continue to be the Sha'alan and their domain is now north-east Jordan and the Syrian Badiya.

22. Rustum, *Al-Mahfuzat*, vol. I, p. 178.

23. Ibid, vol. II, p. 227.

24. Spyridon, 'Annals of Palestine 1821–1841', p. 115.

25. Rustum, *Al-Mahfuzat*, vol. II, p. 443, gives a summary of the document no. 225, File 249, dated 21 Rabi' al-Akhir 1250 H/27 August 1834 AD. A dispatch from Ibrahim Pasha to his father Muhammad 'Ali Pasha; al-Qusus, 'Mudhakkirat', p. 12, has the full text and gives a true picture of the treatment administered to the population of al-Karak after their defeat.

26. Al-Qusus, 'Mudhakkirat', p. 11, confirms that both Shaykh Isma'il Al-Majali, brother of the paramount Shaykh 'Abd al-Qadir, and the latter's son Salih, were captured by Ibrahim Pasha and decapitated in Jerusalem where their bodies were exhibited for three days in the Petra market. Peake, *Tarikh*, p. 175, states that Isma'il, who was paramount shaykh, was hanged in Jerusalem and succeeded by his brother 'Abd al-Qadir.

27. Spyridon, 'Annals', p. 116.

28. Al-Qusus, 'Mudhakkirat', p. 11.

29. Sipano ed., *Mudhakkirat*, p. 77.

30. Spyridon, 'Annals', p. 117.

31. Ibid.

32. Peake, *Tarikh*, p. 170.

33. Rustum, *Al-Mahfuzat*: Ma'oz, *Ottoman Reform*, p. 16.

34. Rustum, *Ara'*, p. 116; Ma'oz, *Ottoman Reform*, p. 14; Bowring,

Report, p. 133: 'There can be no doubt that since the conquest of Syria by Mohamet Ali the agriculture of that country has made considerable progress. Ibrahim Pasha has employed large capital of his own in agricultural pursuits and many villages which under the Sultan's government had been deserted are now again inhabited and their lands cultivated with considerable advantage.'

35. The people of the northern part of Transjordan were only induced to pay taxes after 1851 when the Ottomans made their presence felt through the governorate of 'Ajlun. Similarly the people of al-Salt only paid taxes after 1867 when a governor and a police force were instated among them.

36. 'Aqila Agha, who according to the report of Consul-General Wood had joined the camp of Rashid Pasha with 500 horsemen, was entrusted with the task of enforcing the payment of taxes and fines. Probably it was later in the year that he rode to al-Karak and collected the taxes there. Musil, *Arabia*, vol. III, p. 89, reported that in lieu of all the taxes 'Aqila demanded four sheep from every family. Mansür, *Tarikh*, p. 80, reports that he died in 1870 and was buried in 'Ibilin, a village near Nazareth.

37. Mansür, *Tarihk*, p. 93, 'Aqila was decorated by Napoleon III for his role in 1860.

38. This was brought about by the Ottoman presence in force as of 1867, in al-Salt and 1894 in al-Karak. Governmental authority was so well established in Galilee that 'Aqila's sons left the area completely and resided in the Jordan Valley. However, the 1929 Report of the Legion submitted by its commander, Peake Pasha, to the prime minister mentioned Lieutenant Rida Quwaytin, grandson of 'Aqila Agha, as being then the commander of police in the district of al-Salt. This report is in a file in the Maktaba al-Bahth al-'Ilmi, Ministry of Agriculture, Amman.

39. The original of the letter no. 11 sent by Consul-General Wood to the Hon. Henry Elliot, Britain's ambassador in Constantinople, is dated 27 September 1869. It is now in the archives of the Public Record Office, FO 95/927, Syria and Palestine.

40. Muhammad Rashid Pasha was appointed wali of Syria in 1283 H/1866 and stayed in office, according to the *Salname vilayet-i Suriya* of 1899, for the unexpectedly long period of five years and three months.

41. A new mechanism that was inserted in old rifles to allow the use of ready-made cartridges. As such, it was a great improvement on the old muzzle-loading system.

42. Dhiyab al-'Adwan and his son 'Ali seem to have been taken prisoner and exiled more than once. Peake, *Tarikh*, pp. 170–1, says that Dhiyab returned from exile in Homs in 1841 and that Hula Pasha, governor of Nablus (1863–5), took many of the 'Adwan prisoner and imprisoned 'Ali in Acre for two years until he was released in 1882. Probably Dhiyab was imprisoned by Hulu in 1865 and 'Ali by another governor in 1880.

43. Peake, *Tarikh*, p. 182, relates that Fandi al-Fayiz said after his son

was executed: 'My son and I have been two slaves of the Sultan, now the Sultan has only one.' It has not been possible, however, to confirm this, even from Fandi's grandson 'Awad.

44. While the Bani Sakhr were being punished, the Wuld 'Ali were being rewarded for their support of the Ottoman campaign against the Balqa'. Previously, in the 1830s, the Wuld 'Ali had co-operated closely with Ibrahim Pasha; when Qasim al-Ahmad, the Palestinian leader, and his party sought refuge with them, they did not hesitate to deliver them, against Arab traditions, to the Egyptians who later executed them.

45. Public Record Office FO 95/927. Wood reported that he received news of this battle when his party arrived at al-Zarqa' river (20 kilometres north of Amman). The news confirmed the defeat of Fandi al-Fayiz by Muhammad al-Dukhi, shaykh of the Wuld 'Ali. Wood further commented that 'the greater part of the booty was, it is said, secretly made away with by his tribe.'

46. Al-Hamayda is a tribe that the 1302 H/1884 *Salname vilayet-i Suriya* estimated at 300 tents or 1,500 people. It is also noted that their domain in the Mujib River area is very hard terrain and a crossing can only be accomplished with very great difficulty.

47. There are stories that Dhiyab al-'Adwan co-operated occasionally with the Ottomans, when he needed their assistance. Tales that could not be verified mention that he even once accepted the rank of colonel from them. Sattam Ibn Fayiz co-operated closely with the Ottomans and becoming, according to Salname records, the governor of nayiya al-Giza in 1881, continued to render services to them until his death in 1890. In another district, al-Karak, Tristram, *Moab*, p. 94, mentions that 'the son of shaykh Mudjelli is a Turkish appointed governor, ranks as a Colonel in the Turkish Army, and draws pay as such from the Imperial Treasury, being of course answerable for the taxes due from the district.'

48. Public Record Office, FO 95/927 Syria and Palestine 1869. Consul-General Wood in a despatch to H.M. Embassy in Constantinople mentions that 'al-Salt in 1867 was in the possession of the Adwan beduins, a small but courageous tribe. An inscription records that it was held 150 years by the family of Dhiab al-Adwan their Shaykh.'

49. Public Record Office, FO 78/1978, translation of telegram from Rashid Pasha to the Mutasarrif of Damascus on 10 December 1867, enclosure no. 2 in Dispatch no. 63 from the Consulate-General in Damascus to the British Embassy in Constantinople.

50. Peake, *Tarikh*, p. 264.

51. Al-Qusus, 'Mudhakkirat', p. 60.

Chapter 3

1. At least four of the larger clans in al-Salt are of beduin origin, and so are all the descendants of the semi-settled Balqa' clans during the nineteenth century. In the north, the war cry of the Bani Juhma district is 'Shammar', which may be the result of descent from that famous tribe.

In the south, the whole population is of beduin stock with the exception of al-Karak and the three villages of ʻIraq, Kathraba, and Khanzira.

2. In view of instability and the consequent inability to protect trees, there were no vineyards, olive groves or orchards until the 1920s in the areas east of ʻAjlun, al-Salt and al-Karak. Farhan Abujaber planted fruit trees of all types in an orchard of 16,000 square metres at al-Yaduda around 1900, and although it had a permanent guard it never flourished and only a few standing trees remain.

3. Because of the higher price wheat fetched and the bigger demand for it, farmers preferred to plant more of it. The average at al-Yaduda was near 60 per cent of the whole crop and it is probable that this was the general average for the countryside.

4. A very popular proverb among farming communities was, ʻHe whose harath is not his son, and whose ox is not born by his cow, will have continuous worry.'

5. ʻWheat is with barley the oldest cultivated plant so far discovered and it is with maize the most widespread cereal', *The Atlas of Food Crops*. Until the 1940s wheat continued to be in the Transjordanian countryside the natural unit for measurement. It was thus the main item in barter transactions.

6. This record in one of the account books of al-Yaduda also gives the dates and quantities that appear later in this chapter.

7. This was confirmed by Mr Husni Shuman Ahmad Sawbar, during an interview held on 15 April 1985. Mr Sawbar, who was born in 1919, has for many years been mayor of Wadi al-Sir.

8. Other than kirsanih, there were three varieties of vetches: *jilbana*, *suʻaysa* and *niʻmanih*.

9. From the account books at al-Yaduda.

10. Abu Rumayla, *Al-aʻshab*.

11. Ibid.

12. Economic conditions in general were hard in Transjordan during the nineteenth century and the first four decades of the twentieth. Ordinary farmers who were running their own faddans were unable to pay any cash wages. It was only the landlords of larger estates who could afford this extra expense before the baydar.

13. The term *hay* is adopted for the ear end of the cereal stalk which is used for fodder.

14. Peake, *Tarikh*, p. 60. The Ruwala tribe, who controlled the Jawf area until the beginning of the twentieth century, used to graze their herds in Transjordan. As a result they and the Wuld ʻAli had many fights with the Bani Sakhr.

15. This is an expression that applied to the nomads of Transjordan, especially the Bani Sakhr in al-Balqa' and the Sardiya alliance in the north.

16. Malaria in the Jordan Valley continued to be a problem for the Jordanians until the middle of the twentieth century. It was eradicated after an intensive government drive and the dedicated work of the medical officers, Dr Tannus Qawar in the middle part and Dr Saʻd

Nasrallah in the northern part.

17. Gypsies in Transjordan were divided into two groups, the craftsmen and the entertainers. Both were looked down upon by the general public and the saying 'He is a Nawari (gypsy)' is still used to describe a man without character.

Chapter 4

1. From the west came Palestinians who were either unhappy with the Ottoman conscription policy or were looking forward to a better life in the virgin lands of Transjordan. Generally they came from Jabal Nablus, but there were also those from Jabal al-Khalil (Hebron) and Jabal al-Quds (Jerusalem). The migrations from the south were either beduin movements from the Hijaz and Najd or from among the population of southern Transjordan who left because of instability in their areas. A study of Peake's list of clans in *Tarikh* reveals the large number of clans and families and their different origins.
2. Peake, *Tarikh*, p. 298.
3. Burckhardt, *Travels*, p. 266.
4. Peake, *Tarikh*, p. 169.
5. Ibid., p. 168.
6. Ibid., p. 169.
7. Seetzen, *Brief Account*, p. 37.
8. These 'Adwan slaves are to be differentiated from the al-Juhran clan who were fighting slaves in the 'Adwan camp. They made such a good name for themselves as fighters that a proverb, still in use, says: 'Al-Sit lil 'Adwan wa al-fi'l lil Juhran,' meaning 'The renown is for al-'Adwan when the actual fight is carried by al-Juhran.'
9. Burckhardt, *Travels*, p. 350; Buckingham, *Travels*, p. 34.
10. Burckhardt, *Travels*, p. 355.
11. Ibid.
12. Ibid., p. 354.
13. A member of a Salti clan. He kindly related these stories on 4 April 1981, when he was 70 years old.
14. A member of another Salti clan. He kindly gave his assistance, when he was still a high official at the Lands and Survey Department in Amman, on 3 August 1984. He retired in 1985 at the age of 65.
15. On 6 July 1812, Burckhardt, *Travels*, p. 356, wrote the following: 'I rode over to Feheis where the greater part of Szaltese were encamped for the labours of the harvest.'
16. Until the 1950s both these tribes used to spend the winter in encampments in the Jordan Valley and then move in early May to the plateau between al-Salt and Amman.
17. As usual in these cases, there is no agreement on the names of these four clans. Further research on the subject will not be very useful because of the nature of the subject and the fact that there are very few elderly people still alive who may remember having heard about such matters.

18. These Christian clans allotted lands were the Qawaqsha, Fawakhriya, Dababna, Qa'wara, Nubur, Hatatra, Haddadin, Suwaysat, Makhamra, 'Akarsha, Ziadat and Sumayrat. The last five moved to al-Fuhays just before the turn of the century and reside there still.
19. This name was already in use during Burckhardt's visit in 1812 (*Travels*, p. 349). There are different stories about the origin of the name. The one that seems to be most popular is that a group of Kurdish pilgrims camped in it for a while on their return from the holy places.
20. The stories are many about the Salti fights with their beduin neighbours, the forces of Dahir al-'Umar and the forces of the walis of Damascus. Burckhardt, *Travels*, p. 349, wrote about al-Salt: 'its inhabitants are quite independent. The pashas of Damascus have several times endeavoured in vain to subdue them.' Sixty-five years later they were, however, subdued by Rashid Pasha when he entered the town in 1867, at the head of an Ottoman force.
21. Announcements for sale still appear in daily papers every now and then in which the words used concerning the offered property are: 'itsalat ilayhi bi al-qisma al-'asha' iriya'; which means, 'it has been attached to him through tribal division [of land].
22. Salman, *Khamsa a'wam*, p. 80. Bishop Salman recorded a list of tribal judges at the beginning of the twentieth century. Among them were: 'Awad ibn Qalab of the Bani Hasan who was known as Qadi al-Balqa'; Sultan ibn 'Ali al-Dhiyab of al-'Adwan; Fawwaz ibn Fayiz of the Bani Sakhr; Salim Abu al-Ghanam of al-Ghunaymat; Shahir ibn Haddid of al-Balqawiya; 'Abd al-Mahdi Abu Wandi of al-'Awazim; Sayil al-Shahwan of al-'Ajarma, Qadar; and Rufayfan al-Majali of al-Karak.
23. Al-Qusus, 'Mudhakkirat', p. 27.
24. Peake, *Tarikh*, p. 175.
25. Ibid., p. 25.
26. Al-Zirikli, *'Aman fi 'Amman*, p. 60. Shaykh Mashhur was killed during a mission he carried out at the request of the Transjordan government to reclaim some animals that were plundered from Madaba by a group of the Bani Sakhr. He was educated in Damascus and, for his times, was a most progressive and promising leader.
27. El-Edroos, *The Hashemite Arab Army*, p. 211.
28. Al-Zirikli, *'Aman fi 'Amman*, p. 7.
29. El-Edroos, *The Hashemite Arab Army*, p. 211.
30. Jarvis, *Arab Command*, p. 74.
31. The St George Church in al-Salt, still in use, could have been built in the middle of the nineteenth century. The cemetery, on the other hand, called after Saint Sarah, is definitely much older. A small chapel in it dates back to the early Byzantine period. Burial in the cemetery was stopped towards 1930.
32. There was also an Orthodox community in al-Rumaymin, a few kilometres north of al-Salt, but the church there was built during the twentieth century.
33. It is interesting to note how this tradition, basically beduin in origin, was adopted by the settled communities. Furthermore, giving

land with such largess could have only taken place originally in a society that was not yet deeply attached to agriculture.

34. The whole file of this interesting case was made available for research through the courtesy of the director-general of the Lands and Survey Department in Amman.

35. This information was received during an interview with Shaykh Ghalib Ibn Nawwaf in his house at al-Qastal on 1 September 1985. His cousin Shaykh 'Akif Ibn Mithqal also kindly helped with information.

36. This was mentioned by Shaykh Nawash Ibn Mithqal and Shaykh Ghalib Ibn Nawwaf at al-Qastal on 1 September 1985.

37. The affair, which involved occasional shootings and unpleasant encounters, was sensitive enough not to be mentioned in the family circles. After a few attempts, Sa'id Pasha and his brother Sa'd, who were both present at the Amir's Council in 1924, agreed to talk. The details were revealed by them a few months before Sa'id's death in 1965.

38. Fish, *Bible Lands*, p. 320.

39. Oliphant, *The Land of Gilead*, p. 270.

40. This holding, which was considered large in Transjordan, was sold in 1941 to the Mango family for the equivalent of £40,000.

41. This story was told by Sa'd Abujaber after a riding visit to Falih Abu Junayb, grandson of Rumayh in 1944.

42. Schumacher, *Northern 'Ajlun*, p. 87.

43. The two Lebanese families were al-Murr and Bahuth and the Transjordanian buyer was Shibli Ibn Ibrahim al-Bisharat.

44. The oldest in the Abujaber papers is a receipt signed by 'Ayada and 'Ayyad al-Salim of the 'Ajarma tribe in 1299 H/1881 AD. It confirms the receipt of 62.5 Majidi riyals in addition to what was received by the Tapu Department and the first *sanad* (document) of sale.

Chapter 5

1. It has not been possible to identify one case in Transjordan where farmers were tied to the land or sold with it. What they suffered from was excessive taxation by the walis and mistreatment by the nomadic tribes. Poliak, *Feudalism*, pp. 65f, suggests that as early as the sixteenth century the landlords in Bilad al-Sham were receiving rents in the form of a share from the crop by a *muqasama* system, or division among parties.

2. Firestone, 'Crop-sharing economics', states: 'The Arabic word shadd, or shaddad in its Transjordanian usage, which means *inter alia* the harnessing of animals, was used in the villages for taking on the task of cultivation on a given plot, or on the lands of a given landlord or village.' In most cases the shaddadin who undertook the task provided in full seeds, animals, and labour.

3. Since al-muraba'a required hard and strenuous work it was exclusively undertaken by men. Although women helped in agricultural work in general, it has not been possible to find one confirmed report of

a woman ploughing a field. Likewise al-sharaka was also reserved for men.
4. This term was already in use during the nineteenth century and is mentioned by Abu Yusuf, *Kitab al-kharaj*, p. 91. In the chapter on land rents he states that another type of rent was al-muzara'a, on the basis of a quarter or a third. The term was translated by Firestone as 'co-cultivation'.
5. The information provided by Burckhardt in 1812 is all that is available; although it concerned the Hawran, it also applied to northern Transjordan, which was considered an agricultural and administrative extension of that district. 'Six pairs of oxen were considered a great deal and there were very few who ran an operation of six faddans.' Burckhardt, *Travels*, p. 295.
6. It is interesting to note that the monthly allowance of 48 kilograms of wheat was just sufficient for the needs of a family of five persons; bread consumption would have been at the rate of 10 kilograms per person per month on the average. Also interesting is that flour was not distributed even though it was quite burdensome for people to carry their relatively small rations of whiat every month or two to the flour mills, 10 or 15 kilometres away. It would have been much easier if large quantities were milled and flour was given away in rations, but it seems people wanted to be sure of the quality of their wheat.
7. Burckhardt, *Travels*, p. 297.
8. Probably the difference was due to the fact that security in the Druze area was of a higher standard and therefore the harath did not suffer any losses on account of devastation of crops by beduin flocks of sheep and herds of camels.
9. A ghrara of corn sold for 50 piastres or 2.5 gold pounds sterling. The 10 piastres paid by the harath then were therefore equal to ten shillings.
10. Account book of al-Yaduda for the years 1901 and 1903.
11. Account book of al-Yaduda for 1911.
12. Schaff and Rogers, 'Damascus', p. 190, report that farmers, pressed for taxes, and over-burdened by money-lenders, took refuge in flight and that in this manner the inhabitants of an entire village sometimes disappeared in one night. Although this is most probably an exaggeration, it is clearly indicative that land was still available to those who were willing to farm.
13. It is unfortunate that the account books are in a damaged condition and incomplete. However, it was possible to gather the information for the four years under review.
14. Please refer to chapter 1 above.
15. There were a few of these estates, such as al-Nu'ayma and Sama al-Rusan in the north and al-Yaduda and Hisban in al-Balqa'. The confidence of farmers in the fairness of the conditions regulated by the landlords of these estates encouraged them to disregard their personal interests when determining the wages.
16. Millet (*dhurra*) seems to have been a popular crop until the first

world war, although it was considered inferior and only to be eaten in bad years. Loaves made from millet were called *karadish dhurra* and were cited in proverbs on account of their hard texture and the fact that they were only consumed in days of dire need.

17. It is interesting to note that the landlord's share was a khums (one-fifth), although the basic elements in dry farming were three (land, labour, seeds and animals) or at best four if seeds and animals were considered as two separate items. It seems that the landlords at that time were eager to encourage farming on their lands, and at the same time content to be treated on equal terms with the Ottoman sultan; Canaan, 'The Saqr Beduins of Bisan', confirms that the Saqr were paying 20% of the income derived from the cultivation of the sultan's land as rent.

18. Well water was the most common supply for Transjordanians as the country had only three rivers and a limited number of springs. Every village had its reservoirs which in most cases were old wells that were cleaned and maintained by the new settlers. People had to travel long distances, sometimes over 25 kilometres in one day, to obtain water from springs in years of drought. Therefore the fallahin insisted that their reasonable water requirements were supplied by the landlord as they were not in a position to transport it daily. In later years, it was not uncommon for a shepherd or farmer to buy a well's water from a landlord who had an extra supply.

19. Lambton, *Landlord and Peasant*, p. 306, states that traditionally five elements were taken into account in dividing the crop. These were land, water, draught animals, seed, and labour. In Transjordan, water for irrigation was not available anyway and draught animals and seeds were considered as one element.

20. Sa'id Abjuber used to tell the story of the bad year towards 1900 when most of the people engaged in agriculture in the villages between Amman and Madaba disbanded in mid-March when it was clear that there would be no harvest that year. Many went back to their homes, while families who had no choice but to stay helped in building terraces and clearing wild plants. Generally landlords were under a moral obligation to give these men new agreements and employment in the following year if the shaddad (the assumption of the task of cultivation) was continued.

21. Firestone, 'Crop-sharing economics'.

22. Klein, 'Missionary Tour', p. 125, mentions that he was told: 'about 2,000 yokes of oxen are employed by people of al-Salt and Beduins in this plain [al-Balqa'] for the cultivation of wheat, barley and dhurra.' Agriculture on this scale could have given a yearly crop of some 20,000 tons of cereals. Half of the land under cultivation could have been planted by fallahin.

23. In many cases, the understanding and amity that developed between the landlords and the fallahin continued for generations. At least three generations were involved in farming at al-Yaduda of members of three clans who came from different areas. Al-Batarsa came from Suf near Jarash; al-Badawi came from 'Awarta in Jabal Nablus;

and al-'Awdat came from Sinjil, north of Ramallah. Another clan, al-Qaryuti from Qaryut in Jabal Nablus, were shaddadin for over 40 years at Umm al-'Amad with al-Fayiz, shaykhs of the Bani Sakhr.
24. Because such a large proportion of the population of these two Palestinian villages was working in Transjordan, it was often heard at al-Yaduda, as late as the 1940s, that they were split in half, between Jabal Nablus and the khirab (villages) of al-Balqa'. Al-Dabbagh, *Biladuna Filastin*, part II, vol. II, pp. 395 and 303, gives the following: ''Awarta, 8 km south of Nablus, is considered one area with Awdala nearby. Total area 16,106 dunums of which 5,781 are planted with olives, vineyards, almonds and figs. Total population in 1922 was nearly 1,000 people . . . 'Aqraba, 18 km south-east of Nablus, is considered one area with Fasayil in the Jordan Valley. Total area 190,480 dunums, most of which is barren. It has 4,600 dunums of olive groves and fruit trees. Total population in 1922 was nearly 1,400 people.'
25. Finn, *Stirring Times*, p. 226, states that the peasantry of Palestine were broadly divided into Qaysi and Yamani. Peake, *Tarikh*, pp. 242, 347 and 356, refers the origin of some Transjordanian clans to Jabal Qays, which is also called Jabal al-Khalil.
26. Rarely did any of the tribesmen become a harath on a yearly basis. Generally they undertook the odd jobs, such as watchmen, caravan leaders and helping at harvest time. Most of them were not suited physically to the hard work expected from a harath, but there were also among them a good number who, impoverished as they were, did not deign to become regular farmers.

Chapter 6

1. Ionides, *Report*, p. 139. The following information was given in 1939: The main channel of the river is 20 to 30 metres wide. It is crossed by three bridges at which the sea-levels are −220 metres Jisr al-Majami', −280 metres Jis al-Shaikh Husain and −345 meters at Jisr Allenby.
2. Lynch, *Narrative*, p. 149.
3. Rogers, 'Maritime cities', p. 143.
4. Adam Smith, *Historical Geography*, p. 669.
5. Seetzen, *Brief Account*, p. 14.
6. Thomson, *The Land and the Book*, p. 359.
7. Ibid., p. 360.
8. The full text of these notes is reproduced in the *Palestine Exploration Fund Quarterly Statement* for 1868, pp. 1–9.
9. From an article which appeared on p. 161 of the *Palestine Exploration Fund Quarterly Statement* for 1891.
10. Neil, 'Pits', p. 161.
11. Hill, *With the Beduins*, p. 39.
12. *Palestine Exploration Fund Quarterly Statement*, 1891, p. 199.
13. Ma'oz, *Ottoman Reform*, p. 169, writes that the most neglected public works were roads and their ancillaries — bridges and inns. The

reasons were the corruption of the local authorities and the indifference of the local population.

14. Seetzen, *Brief Account*, p. 38.

15. Al-Jamil, *al-Badu*, p. 252.

16. Bell, *Desert and the Sown*, p. 40, gives an interesting report. She wrote: 'The corn was kept in a deep dry hole cut in the rock and was drawn like so much water in bucketfuls. It had been stored with chaff for its better protection and the first business was to sift it at the well-head.'

17. Oliphant, *Gilead*, p. 270, gives the first reference to this trade. He wrote about Salih Abujaber in 1879: 'He stores his grain away in the large underground vaults which were used for the same purpose in ages gone by and either sells it at Jerusalem transporting it there on his own camels or to travelling merchants who come and buy it of him.'

18. Sa'id Farhan Abujaber used to tell stories about the tribes who measured wheat at al-Yaduda during the years before his birth around 1885. The largest buyers were al-Shararat, followed by the Ruwala and al-Huwaytat. Once, a group came with 40 camels and declined to declare their identity although they were known to be of the Ruwala. After the transaction was concluded, the 40 camels were loaded and the price of the wheat paid in a *surra* (purse) of gold pounds. Sa'id's father, Farhan, suspecting that the load was destined for al-Sha'alan, shaykhs of the Ruwala, instructed one of his men to place the purse in the mouth of one of two sacks to be loaded on the leading camel. When the money was discovered Ibn Sha'alan was very touched and sent a present in return. It was three camel loads of Yemeni coffee and a camel load of cardamoms which was an expensive luxury then as now.

19. Gharayba, *Suriya*, p. 107, mentions that Jerusalem, which had 20,000 people in 1840, was populated by double that number in 1890. Such a number would require nearly 6,000 tons of wheat annually, or the surplus production of some eight villages of al-Yaduda's size.

20. Although the growth of population is generally confirmed, there are differences in the figures. As opposed to those of Gharayba, given above, al-'Arif, *al-mufassal*, pp. 292 and 296, mentions that the population of Jerusalem grew from 20,000 in 1839 to 68,000 in 1861. According to al-Ramini, *Nablus*, p. 16, Nablus grew from 8,000 in 1882 to 17,472 in 1900. According to Seikaly, 'The Arab community of Haifa', p. 20, Haifa grew from 3,000 in 1850 to 9,000 in 1900.

21. Ma'tuq is a Christian family of Jerusalem, of which a part emigrated to Cairo. There is also a Muslim family who originally came from Hebron.

22. Arab mares were in great demand; some were sold for upwards of 500 gold pounds. Money exchange rates varied, but in 1878 Finn, 'The Fallaheen of Palestine', p. 72, mentions that a piastre was nearly 2d. or 120 piastres to the gold pound sterling.

23. As the schools in Transjordan were mainly elementary at the beginning of the century, some of the families found it necessary to send their children to boarding schools in Jerusalem. In 1908 Sa'd and his cousin Salih went to St George's School and two girl cousins went to the

German Schmidt school in Jerusalem.
24. The Nashashibis are an old-established Muslim family, while the Talil are an old-established Christian family in Jerusalem. Salim Marar, the third partner, was a shaykh of the ʻUzayzat tribe in Madaba. His sons were studying at Bishop Gobat's School in the Holy City, and he seems to have had to return to Madaba as a result of the first world war.
25. The remark concerns the bulk of al-Yaduda's yearly labour force, who were generally villagers from the district around Nablus and to a lesser extent from the district around Jerusalem. They all went home at the end of al-baydar (threshing) in August or September. They preferred to receive their pay in Jerusalem rather than at al-Yaduda, for fear of robbers on the roads.
26. Tattan, as I was told by Dr Da'ud al-Husayni in Amman on 10 November 1985, was originally from Hebron where his family is known by the name al-Mani.
27. These four belonged to known families in Jerusalem. According to a death announcement found among the documents, Batatu died on 11 May 1916 in Damascus at the age of 56, after acquiring from Shaykh Rumayh Abu Junayb 17,000 dunums in al-Tunayb east of al-Yaduda. This estate was sold to the Mango family in Amman in 1941 for about 40,000 Palestinian pounds, equivalent to 40,000 pounds sterling.
28. It is not clear why the Municipal Council of Jerusalem wanted this quantity of wheat or how Farhan gave them the undertaking to deliver it. Although al-Budayri's letter is written on an ordinary piece of paper, like the receipt, it appears to be an official document. Possibly the wheat was required for distribution by the council to the poor families in Jerusalem.
29. The patriarchates in Jerusalem normally required hard wheat to turn into burghul. This is boiled wheat dried in the sun, crushed and cooked as rice or used as an ingredient for different dishes, especially kubba when it is mixed with pounded meat. Farhan was a follower of the Latin patriarchate of Jerusalem until 1878 when his father Salih, as a result of disagreement, became a follower of the Orthodox patriarchate.
30. Letter from Farhan to his son Saʻid at al-Yaduda, 26 July 1913, and letters from Tattan to Farah at al-Yaduda dated 17 July 1912 and 21 Shaʻban 1330 H/1912 AD.
31. Letter from Farhan to Saʻid at al-Yaduda, 25 July 1913, and letter from ʻAbdallah Badr to Farhan at al-Yaduda, 12 August 1914. Most probably the banks stopped allowing credit to anybody after 2 August 1914 when Turkey declared mobilization. The official declaration of war was made on 5 November 1914.
32. Conditions do not seem to have changed much after the opening of the Hijaz railway between Damascus and Madina. Transport was mainly needed to Jerusalem and Nablus. Rates per ton were competitive with rates for animal transport: a circular found at al-Yaduda (undated, but probably 1920) gives the fare per ton between Amman and Damascus as £1.814 and between Amman and Maʻan as £2.332.
33. Since Amman had the advantage of having a railway station, it is

probable that its merchants used the Hijaz railway for the transport of cereals to the Palestinian cities, although this could have been more expensive than animal transport.

Chapter 7

1. This was his second posting in Damascus. The first was for one year and four months which started in 1292 H/1875 AD.The second lasted for five years and fourteen days and started in 1296 H/1879 AD, immediately after the governor hip of the renowned Midhat Pasha. An ordinary writing copy-book, found among the papers in Karton no. 33, Zarf no. 93, Kissim no. 18 in the Yildiz Esas Evrak at the Başbakanlik in Istanbul, turned out to be the draft of a report submitted to the governor of Damascus on 28 Tammuz 1296/1879 AD. It contains complete lists of names, dates of start of work and monthly salaries of all employees in the wilaya of Damascus. The comprehensive and original information contained in its 54 pages is referred to in this study.
2. Sanjaq Balqa' had Nablus as its administrative centre and, in addition to Jabal Nablus, covered the territory between the Zarqa' and Mujib rivers in Transjordan. Al-Salt was the seat of the qa'immaqam who administered the Transjordanian part of the sanjaq.
3. The report mentioned only two nahiyat in the whole of Transjordan, Ma'an and Jiza. Each had a mudir nahiya as administrator but only Jiza had a *katib nahiya*, or scribe.
4. Both belonged to well-known Lebanese families. During their stay in Transjordan they acquired agricultural lands and houses. A good part of the lands in Zabda, a village a few kilometres south of Irbid, is still the property of the descendants of Farkuh. The descendants of Kassab lived in Beirut and disposed of some property in the 1960s.
5. Salname Vilayet-i Suriya, 1288 H (1871/2 AD), pp. 90, 91.
6. Ibid., p. 91.
7. Salname Vilayet-i Suriya 1290 H (1873/4 AD), p. 101.
8. *A Survey of Palestine*, p. 246.
9. Three books were consulted and there was no agreement among them: Kurd 'Ali, *Khitat al-Sham*, p. 193; al-Hasani, *Tarikh*, p. 230; 'Awad, *al-Idara*, p. 168. Himadeh, 'Taxation', also provides different figures and dates.
10. Himadeh, 'Taxation', p. 187.
11. A *Survey of Palestine*, p. 246.
12. Tute, *Ottoman Land Laws*. pp. 1, 3, 7, 10, 14, 15.
13. Wavell, *The Palestine Campaigns*, pp. 185, 186. Al-Salt was captured on 29 April 1918, and British forces had to evacuate the town, on the orders of General Allenby, during the night of 3/4 May 1918, to avoid being beseiged by the advancing Turks. The people of al-Salt who followed the British forces out of fear of fighting and disturbances continued their journey to Jerusalem where some of them stayed for over six months. The villagers of Abu Dis, east of Jerusalem, led by the

'Uraykat clan, gave real assistance and extended Arab hospitality to the refugees.

14. 'Awad, *al-Idara*, p. 176. A receipt among al-Yaduda papers dated 20 March 1332 H/1914 AD states that Farhan Abujaber had paid 4,105 piastres as aghnam tax for the two preceding years on 458 sheep and 55 goats, a total of 513 at 4 piastres each per year.

15. Oliphant, *Land of Gilead*, p. 271. 'The Arabs, although they cultivate the land, pay nothing on their crops, and the government loses therefore the entire revenue from this source if there were a settled population.'

16. Schumacher, *The Jaulan*, p. 54.

17. Musil, *Arabia*, pp. 90, 91.

18. Mantran, *Règlements fiscaux*, p. 39.

19. The Hijaz railway was a major Ottoman and Muslim enterprise; the aim was to construct a railway between Damascus and the Hijaz, a distance of over 800 miles. Work started on it in 1900 and was completed in 1908 when the line was officially inaugurated after reaching Madina. In Transjordan, the line and its property has the status of a *waqf* (endowment).

20. Oliphant, *Land of Gilead*, p. 271.

21. The al-'Ajarma receipt mentions no amount and merely states that Nayf Mustafa al-Shahwan al-'Ajarma should receive badal 'ashar the mahla fields until the year 1316 H/1898 AD. It is witnessed by Ibrahim Bisharat, founder of the well-known Bisharat clan, and Yusuf 'Asfur, first merchant in Amman then, and grandfather of the well-known 'Asfur family in Amman. The al-Fayiz receipt is dated 13 Rajab 1325 H/1907 AD. It states that it was for the 'ashar of al-Nu'ajiya and has the names of six people as witnesses.

22. Al-Qusus, 'Mudhakkirat', p. 56.

23. A clan of the Karadsha tribe in Madaba.

24. Shaykh of al-'Uzayzat tribe and also of Madaba as a whole.

25. A family in al-Salt which was originally of the 'Uzayzat tribe in Madaba.

26. Farik is roasted green wheat and is cooked and eaten like rice.

27. An interesting paper was found with the documents of this case in the Abujaber collection. This is a draft copy in Turkish with a translation in Arabic which was probably referred to by the Abujabers when presenting their case. It is dated 21 December 1330 H/1910 AD, and concerns the taxes of al-Juwayda owed by the sons of Ibrahim Abujaber. The multazim is the same, 'Isa Hamarna of Madaba.

28. Batna is a settlement in the district of al-Salt. The document was among al-Yaduda papers because the Abujabers probably referred to its contents when presenting their case to the government.

29. Saydu is a Kurd who came to Amman during the early part of the twentieth century and succeeded in establishing himself as one of the richest landowners in the country. The family is now well known and his son Husni is chairman of the Jordan Bank. His partners were Sadiq al-Batikhi, a prominent merchant of Damascene origin, and Mahmud Sa'id

Amin, probably a Circassian of Amman's Muhajirin quarter.

30. It was not unusual for tax-farmers, who were so unpopular, to use flowery language and courteous manners in their dealings with their clients in influential circles. They even extended loans when necessary, as is shown by a letter in which Mashhur Ibn Fayiz of the Bani Sakhr asked Sa'id Abujaber kindly to settle the barley he owed him (eleven camel loads) directly with Saydu, from whom Mashhur had received a loan of barley. This letter is dated 12 August 1233 H/1915 AD and the delivery was made against Saydu's signature two days later.

31. Ismai'il Bey was also a Circassian of the Muhajirin quarter in Amman. The title bey probably meant that he worked with the Circassian contingent in the service of wilaya of Damascus, especially in the area of al-Karak.

32. Both Zizia and Amman had a strong Ottoman presence because they were both important stations on the Hijaz railway. Zizia was, however, in the domain of the Bani Sakhr tribe.

33. Although al-Yaduda is only 10 kilometres from Amman, the cereals were delivered to Zizia which is nearly 20 kilometres away. This may be explained by the fact that it was easier to hire camels from the Bani Sakhr for the transfer of the cereals to Zizia because it was on their route to the eastern domain. Transportation of the whole quantity required between 450 and 500 camel trips to accomplish.

34. The letter was found among the Abujaber papers at al-Yaduda in 1963. My emphasis.

35. Sijill 902 at the Başbakanlik in Istanbul, containing items of correspondence with wilayat Suriya. The order dispatched to the governor in Damascus is dated 17 Rajab 1288 H/13 July 1871 and reads: 'Sa'id Pasha Mustasarrif al-Balqa' received complaints from Istanbul directed against him and the tax-farmers. These complaints are lodged against the misuse of *nizam al-iltizam* [the system of tax-farming]. Inquiry is to be made into this matter.' It is important to mention in this connection that 'al-Karak and Tafila revolted during 1911 because of dissatisfaction with tax assessments and complaints against similar misuse in the process of tax collection. The revolt was put down by the Ottomans after serious loss of life and damage to property, especially government buildings and records.

36. Proceedings of the Council of the Governorate of Damascus, Register no. 5, p. 363, dated 5 Shawwal 1261 H/1845 AD.

37. A search was made in the archives of the Ministry of Finance in Amman, through the co-operation of the director general of the Ministry, but no records were discovered for the period under research. There is still hope of finding some records about the 'Ajlun area, where a governor was appointed in 1851 and the district was attached to Hawran. The records in al-Salt were completely destroyed by a fire that demolished the *serai* (government house) in 1935.

38. It is to be remembered that the Ottomans were only able to take control of the southern area when an expeditionary force took control of al-Karak in 1894.

39. Schaff and Rogers, 'Damascus'.
40. Oliphant, *Land of Gilead*, p. 129.
41. 'Awad, *al-Idara*, p. 176, mentions that around 1876 this animal tax was 8 piastres for every sheep or goat and 20 piastres for every camel.
42. Oliphant, *Land of Gilead*, p. 121. Probably the author's information involved the faction of the Bani Sakhr tribe in the 'Ajlun Qada'. In al-Balqa' some of the Bani Sakhr clans were already paying taxes on their animals in 1879 since their Shaykh Sattam was by then governor of Zizia and co-operating with the authorities.
43. Oliphant, *Land of Gilead*, p. 203.
44. Public Records Office, FO 195/675, British consul Finn to Rt Hon. Earl Russell, dated 21 October 1861 in Jerusalem. The letter also mentions that 'at the more wealthy town beyond al-Karak, called Tafila, the inhabitants have abandoned their place'.
45. Peake, *Tarikh*, p. 170.
46. The governor did not succeed in collecting one single piastre as the 'Adwan had over 200 horsemen to his 50, and intimidated him by their aggressive behaviour. Sa'd Abujaber told the story of the younger 'Adwans throwing small stones at the Pasha's party with the insulting remark: *Hayy Bawlu pasha* (Hey so you come with Bawlu Pasha). Bawl in Arabic means urine and the insulting nature of such remarks is therefore very clear. A year later the governor avenged himself and took 'Ali al-Dhiyab and a group of his followers captive and imprisoned them for nearly two years in the gaol at Acre.
47. Musil, *Arabia*, vol. III, p. 89.
48. Ibid. These were the paramount shaykhs Salih, Khalil Ibn Mustafa and Faris Ibn Salama.
49. Ibid., p. 91.
50. The oldest tax receipts found in al-Yaduda's papers involving payment of money instead of cereals were those of Sa'id and Sa'd, sons of Farhan Abujaber, for the year 1926. The tax for one-third of al-Yaduda amounted to 152 Palestinian pounds and 976 mils. During that period Transjordan was still using the Palestinian currency.
51. The last revolt in al-Karak against mistreatment of farmers and corruption in the collection of taxes happened in 1911. 'Awda al-Qusus renders a detailed account of it in his unpublished memoirs.

Chapter 8

1. Fadiya, youngest daughter of Salih Abujaber, born 1875, told me that her ancestors travelled a lot. They moved from the southern part of Transjordan to the Hawran; then to Marj'ayun in southern Lebanon where they left cousins, Awlad Nayifa; then to Nazareth, Nablus, al-Salt, and last to al-Yaduda.

There have been very friendly relations between the Abujabers of al-Salt who are frequently called al-Jawabra, and the 'Awran, the leading family of al-Jawabra of al-Tafila. The relationship was even considered as that between cousins. The coincidence of having the same name could mean that the Abujabers of Nazareth and al-Salt may have been a branch of the clans in al-Tafila. The war-cry is *Subyan al-Jawabra* (the brave Abujabers), and this is considered as another indication of the old relationship.

2. Mansür, *Tarikh al-Nasira*, p. 205. The sale is probably registered in the Nazareth Sijill al-Fahumi (Court Register). Mansür does not specifically mention it. Records in the Latin church in Nazareth mention baptisms, marriages and deaths of Abujabers during the early decades of the eighteenth century. Although there are still a few Abujaber families in Nazareth, a much larger group has migrated to the Americas.

3. Church Missionary Society, *Register of Missionaries*, p. 324. The Reverend Michael Kawar was ordained in 1871 and died in 1886.

4. The Salname Vilayet-i Suriya 1285 H/1868 AD mentions that there was in al-Salt a Majlis Idara and a Majlis Da'awi. Salih Abujaber Effendi was one of five members of the latter.

5. A good account of Salih's life, activities and standing is given by Medebielle, *Salt*.

6. Salname Vilayet-i Suriya for 1886.

7. Abu Wandi must have been a very progressive shaykh. The report of proceedings of the Oriental Scientific Assembly held in Beirut during 1882 gives the following information concerning cultural activity in al-Salt: 'There is but one school in the area neighbouring al-Salt and among the beduins. It is at the Balqa' encampment of al-'Awazim and run at the expense of their Shaykh Abu Wandi. It has 20 students and a Khatib [teacher] who teaches the children.'

8. Burckhardt, *Travels*, p. 367.

9. Ibid., p. 354.

10. Partington, *Alkali Industry*, p. 61.

11. Burckhardt, *Travels*, p. 351.

12. Karpat, 'Social stratification', p. 97; Taqi al-Din, *Muntakhabat*: the author mentions that some of his family in later periods became involved in trade or became interested in agriculture, as did others among the dignitaries of Damascus.

13. The word khirba (plural khirab) means in Arabic a ruined site surrounded by fields. It was used by settled folk and beduins alike. In the middle of each of these deserted villages there was a mound covered with ruins attesting to the fact that they had formerly been prosperous and well populated.

14. Jaussen, *Coutumes*, pp. 175, 244.

15. Al-Dabbagh, *Biladuna*, vol. IV, part II, p. 512.

16. Savignac, 'Une église byzantine', p. 434.

17. Burckhardt, *Travels*, pp. 352, 357.

18. Salname Vilayet-i Suriya for 1871 mentions al-Yaduda among the villages of Qada' al-Salt.

19. Waterfield, *Layard*, p. 36.

20. The Bani Sakhr clans, during their supremacy in the northern part, acquired rights in different villages. Mujhim Ibn Haditha al-Khuraysha,

as governor of al-Salt, reported on 6 July 1985 that his tribe had lands in Fa‘, Burayqa, al-Khanisiri on the eastern side of Irbid, and Maru and Kufr Jayiz north of Irbid.

21. These happenings were related to me on 1 August 1983 by Na‘ur Ibn Falih Ibn al-Sayyid Ibn Rumayh Abu Junayb, having himself heard them from his father.

22. Al-‘Uzayzi, *Qamus*, p. 21.

23. Although not one of the bigger tribes in Arabia and Syria, the Bani Sakhr had a haughty attitude that was probably the result of their horsemen's performance in the field. The poet was telling them to be more considerate and to accept that other tribes have valour and horsemanship. Parts of this qasida are related by Musil, *Arabia*, vol. II, p. 239, and by al-Sudayri, *Abtal*, p. 255. As there are differences between the texts, the free translation is from the Musil version. It is the older one, and was most probably derived from the nearest of kin to the poet, in view of Musil's very close relation with the princely families of the Ruwala.

24. The story of this important agreement, which led to the family's prominence in the field of agriculture and other areas, was naturally a subject of common discussion in family circles. I heard about it on many occasions from my great-aunts, my uncle Sa‘id, who died in 1965 at the age of 80, and my father Sa‘d, who died in 1976 also at 80.

25. The great plague that swept over Bilad al-Sham in 747 H/1347 AD and the two successive waves of 790 H/1388 AD and 833 H/1429 AD. Further information on this subject may be found in Ghawanma, 'Al-ta‘un wa-al-jafaf'.

26. Hütteroth and Abdulfattah, *Historical Geography*.

27. Jawlan supplied oxen to the neighbouring areas because settlers there had the pasture lands and water resources to raise cattle on a relatively large scale. The writer remembers that the Jawlani cattle were famous in al-Balqa' as recently as 1940, when oxen of the 'black and white' (*abraq*) species were still being imported for reproduction and farming purposes.

28. It is not certain who first imported a tractor into Transjordan but it is known that Hanna al-Bisharat and Husni Saydu al-Kurdi imported small tractors and equipment in the 1920s. Mithqal Pasha al-Fayiz and the Mango family imported caterpillar tractors in the early 1940s.

29. Manual work was looked down upon by the nomadic tribes and the well-remembered expression of beduin shaykhs *Ikhs yal-fallah* (shame on thee, farmer) attests clearly to this feeling among them, which remained prevalent until the 1940s and 1950s.

30. Salman, *Khamsa a‘wam*, p. 165.

31. Musil, *Arabia*, vol. II, p. 89, mentions that Salih had taken over the iltizam for collection of taxes in al-Karak in 1878. Later, in 1894, he played an important role in the peaceful entry of an Ottoman expeditionary force to al-Karak and became the contractor for the supply of provisions to the force. He made good money which he spent on building his big house in al-Salt, the largest in the city up till now.

32. Nawfal, *al-Dustur*, p. 883. The charter for the establishment of the Ottoman Bank was issued in Istanbul on 14 February 1863. Although the bank opened branches in Palestine during the 1880s it only opened a branch in Amman in 1928.
33. Jarida al-da'awa wa-al-sukuk al-shar'iya al-waqi'a fi mahkama al-Salt al-shar'iya as of the start of Rabi'al-Akhir 1305 H/1886 AD (record of claims and agreements in al-Salt's Religious Court).
34. The atmosphere of detachment was undoubtedly also brought about by tribal considerations. Al-Yaduda was considered part of the Bani Sakhr dira while al-Juwayda was part of al-Balqa'. Each party had to operate within the requirements and dictates of these two alliances.
35. Chicken was considered a delicacy in Transjordan until the 1960s. Prior to that it was difficult to raise them in big numbers due to frequent waves of pestilence.
36. The Circassians in Transjordan realized within a short while after their arrival that it was far more economical to slaughter lambs instead of horses, especially since they no longer had the herds they had previously owned in the Caucasus.
37. This could have been the case until 1920 after which names of workmen from only the Nablus area appear in the registers.
38. The fields in Umm al-Basatin could not be joined to al-Yaduda because administratively it is part of the province of Madaba while al-Yaduda is part of the province of Amman.
39. Buckingham, *The Arab Tribes*, p. 33.
40. The term *baylik*, according to Muhammad al-Rashid (13 January 1985), was a Turkish word used in al-Yaduda after the fashion in the area south of Damascus, and meant the household. Faroghi, *Towns*, p. 342, mentions baylik as being the quality of a bey (lord, prince) and the usage in southern Syria probably derived from that.
41. The private dwelling was a two-storey building built over an old Roman structure which was used as a store. The higher part of the building collapsed during the 1927 earthquake but the Roman structure is still intact and in use.
42. The large estates around Damascus and in al-Biqa' valley were called *sawamat* and were referred to as such in records. Sipano, *Tarikh*, pp. 42, 68, mentions that in 1808 Yusuf Kanj Pasha moved to confiscate the sawamat of al-Biqa' controlled by Amir Bashir and the son of Junblatt and that they were vast estates yielding very large incomes. The same name is given to the farms in al-Biqa' during the wilaya (governorship) in Damascus of Darwish Pasha in 1819. It was also used for the estates of al-Fa'ur princes of al-Fadl tribe in al-Jawlan.
43. Medebielle, *Salt*, p. 11.
44. Oliphant, *Gilead*, p. 270.
45. Muhammad al-Rashid, in an interview on 30 October 1984, used the term *fidawiya* for this group which means in general 'bodyguards'. Literally translated it means those who are prepared to sacrifice their lives. Whenever an Abujaber rode on an errand or for a visit, one or more of them usually rode with him.

46. The word *shukara* originally derives from the verb shakara, meaning ‘he thanked’. The usage confirms the original character of the gesture as a gift, not wages.
47. Many of these were men who stayed in the country after having performed the pilgrimage to the holy places.
48. A few entries have been found in the accounts of al-Yaduda regarding purchases of coal from people in ‘Ajlun and Sawf near Jarash.
49. Like the Abujabers, none of them live now in al-Salt. They have also sold parts of the estate, but the third and fourth generations still own most of the village of Umm al-Kundum.
50. Frayh was the first to marry into the Qa‘wars, who claim to be descendants of the Ghassanids. Later Sa‘id and Sa‘d also married into the family.
51. The first generation who came to al-Salt as merchants acquired Hisban. Among the second generation, Sulayman became prime minister of Jordan in 1956.
52. Sa‘id, the son of Khayr, was mayor of Amman when Amir Abdullah arrived there from al-Hijaz in 1921.
53. Al-Sharabi was a merchant from Nablus who settled in al-Salt.
54. Al-Fayyad is a Salti of the Qutayshat clan. He acquired many fields before the turn of the century but sold them during the world depression in the 1930s.

Chapter 9

1. Schumacher, *Northern Ajlun*, p. 28.
2. Tristram, *Israel*, p. 465.
3. Schumacher, *Northern Ajlun*, p. 28.
4. Schaff and Rogers, ‘Damascus’, p. 190.
5. Peake, *Tarikh*, p. 170. A well-remembered expedition is that of Hulu Pasha al-‘Abid, governor of Nablus, 1863–5, who on failing to collect any taxes avenged himself by taking Dhiyab, shaykh of al-‘Adwan, prisoner together with many of his tribesmen.
6. Schumacher, *Northern Ajlun*, p. 29.
7. Al-Madi and Musa, *‘Al-Urdunn*, p. 7.
8. Peake, *Tarikh*, p. 276.
9. Schumacher, *Northern Ajlun*, p. 130.
10. Ibid., p. 139.
11. Schumacher, *The Jaulan*, p. 52.
12. Hütteroth and Abdulfattah, *Historical Geography*, p. 205.
13. Burckhardt, *Travels*, p. 268.
14. Land area figures are given as recorded in the List of Surveyed Sites prepared by the Department of Lands and Survey in Amman and published in their Annual Report for 1975.
15. I am indebted for this detailed information to Mr Na‘im Abu al-Sha‘ar of the Numura clan. The interview was held on 13 April 1985.
16. Peake, *Tarikh*, p. 289.

17. Due to lack of documents, the role of Dhahir al-'Umar in northern Transjordan during the eighteenth century is not yet written. Mention of it is made by Mu'ammar, *Dahir al-'Umar*, p. 274.
18. Peake, *Tarikh*, p. 281.
19. Ibid., p. 282.
20. This document, though mentioned by Peake, *Tarikh*, p. 282 (written around 1930), has not yet been published. It was kindly provided by Mr 'Ali Mansur Wanas al-Hindawi, great-grandson of Hindawi, during an interview in Amman on 6 April 1985. The translation was kindly checked by his Lordship Fayiq Halazun, retired Judge of the High Court, Amman.
21. Farwell, *A Biography*, p. 270: 'Damascus is probably the most fanatical town in the Ottoman Empire', wrote Sir Henry Elliot, British ambassador at Constantinople (1869) to the Foreign Secretary, the Earl of Clarendon.
22. Finn, *Stirring Times*, vol. II, pp. 424, 440. Writing about the religious uprising in Nablus on 4 April 1856, Finn, consul-general of Britain in Jerusalem 1845–63, stated that: 'The cup of fanaticism was full and the one drop more caused it to run over. The movement was from first to last anti-Christian for which the incidents above described were but the pretext.'
23. The Salname-yi devlet-i aliyye of 1285 H/1868 AD mentions Muhammad Sa'id Bey as the mutasarrif of Liwa' al-Balqa', of which the capital was Nablus.
24. This information was kindly provided by Sa'd al-Na'im al-Murrayan during an interview held in al-Nu'ayma on 9 April 1985.
25. Much information was derived during a few hours' interview with Shaykh Mansur Wanas al-Hindawi at his house in al-Nu'ayma on 9 April 1985.
26. Information regarding the area of the holding by the descendants of al-Hindawi was kindly given me by Mr 'Ali Mansur Wanas al-Hindawi in Amman on 29 May 1986. He confirmed that during the land survey of the early 1940s their quarter was nearly 12,000 dunums. It was divided in equal shares between three sons, Wanas, Muhammad and Salim at Hindawi's death in 1320 H/1902 AD.
27. Qusus, 'Mudhakkirat', p. 30.
28. The price of land was very depressed during these years and a dunum in that area could have been sold for 5 shillings at the most.
29. In accordance with the figures recorded by the Department of General Statistics during the census held on 10 November 1979.
30. Most of the Palestinian refugees in the area live in al-Husn camp which, according to the census of 10 November 1979, had a total of 8,792 persons.
31. Please refer to chapter 2 on the history of Transjordan during the nineteenth century.
32. Khanasiri has an area of 27,671 dunums while Fa' has an area of 18,322 dunums.
33. These are: Hawsha 40,703 dunums; Burayqa 29,912 dunums;

Buwayda 21,882 dunums.
34. The youngest of Hindawi's sons, Salim Pasha, died in 1951. He was born at al-Nu'ayma in 1885.

Chapter 10

1. Sattam was born around 1830 and died in 1890. He became shaykh mashayikh Bani Sakhr in September 1881, and governor of Zizia (Jiza) under the Ottomans until his death. For more information see Tristram, *Moab*; Conder, *Heth*.
2. Fandi was born probably around 1805 and died in 1878 and was buried in the 'Adwan territory. Fish, *Bible Lands*, confirms that he was not quite 70 when he visited him in 1875.
3. Peake, *Tarikh*, pp. 270 and 219.
4. 'Awad was born in 1891 four months after his father's death in 1890. His memory was outstanding. He lives in Sahab. He is also known as Shaykh 'Awad Ibn Fayiz and Shaykh 'Awad al-Sattam.
5. Ghalib is the son of Nawwaf, Sattam's only son from 'Alya al-'Adwan. He was born in 1911 and lives in al-Qastal. He is also known as Shaykh Ghalib Ibn Nawwaf and Shaykh Ghalib Ibn Nawwaf al-Sattam.
6. Ghalib related during the interview how his father, Nawwaf, gave all his lands in Khirba Zuwayzia to al-Duraybi clan as dowry for three of their daughters whom he married one after the other during a short period. He also mentioned that a part of Khirba Barazayn was given as dowry to 'Ayda.
7. This is according to the story told by Shaykh 'Awad al-Sattam on 1 August 1985.
8. 'Ali was born around 1835 and died around the turn of the century. He assumed the post of shaykh mashayikh al-Balqa' during the lifetime of his father Dhiyab, who lived to be 100.
9. Peake, *Tarikh*, pp. 171, 172.
10. These details were kindly given by Shaykh 'Awad al-Sattam.
11. It is most difficult to obtain authentic confirmation of such matters in a beduin society as people are apt to forget or mix up names after the third or fourth generation. As far as I could gather, there is no comprehensive genealogy for the Bani Sakhr or even the clan of al-Fayiz at the present time. Conder, *Survey*, p. 295 mentions that Fandi had eight sons; Shuwayhat, *Al-'Uzayzat*, p. 75, gives the number as twelve, which was confirmed by Shaykh 'Awad and by a list I found in the family papers written probably by my father, Sa'd Abujaber, some 60 years ago.
12. Shuwayhat, *Al-'Uzayzat*, p. 75.
13. Tristram, *Moab*, p. 344.
14. Al-hadhiya is 'a gift asked from he who made a gain or acquired one', Qamus, *al-'Uzayzat*, vol. I, p. 208.
15. Tristram, *Moab*, p. 281.
16. Shuwayhat, *Al-'Uzayzat*, p. 76.

17. Evidently Sattam did not succeed in winning over the Zabn clan in spite of his largess. Peake, *Tarikh*, p. 219, mentions that Shaykh Minwir Ibn Zabn declined to acknowledge Sattam's chieftainship and for a time represented his clan in person in Istanbul. Shaykh Turad Ibn Zabn, famous for his raids against the Ruwala, who died around 1900, never recognized the al-Fayiz as paramount shaykhs of the Bani Sakhr.
18. Bell, *Desert and the Sown*, p. 34.
19. A few years earlier a similar large-scale acquisition of land took place when Ibn Muhayd of the Fad'an tribe acquired 20 villages in the Aleppo area; Chatty, 'Tawasu' al-badw'.
20. Ponsonby, *Handbook*, p. 57.
21. Tristram, *Moab*, p. 153.
22. Ibid., p. 344.
23. Schumacher, *Jaulan*, p. 51: 'Thanks to the vigorous action of the Turkish authorities during the last 30 years, the thievishness and annoyance of the Bedawin has been put a stop to, successfully. The fighting tribes were threatened with extermination which was in part actually effected.'
24. Letter no. 11 sent by the British consul-general in Damascus to the British ambassador in Constantinople (FO 95/927 Syria and Palestine, Public Records Office, London) states that Rashid Pasha, governor of wilaya Suriya in May 1869, sent with the advance force against the Bani Sakhr and 'Adwan a battalion of infantry on dromedaries carrying Snider rifles.
25. 'Rifle', *Encyclopaedia Britannica*, vols 23–4, 11th edn, p. 326; Dudik, *Firearms*, pp. 9 and 65; Wilkinson, *Great Guns*, p. 158.
26. Tristram, *Moab*, p. 158, enumerates the advance guard under the Pasha of Nablus that went out to rescue them from alleged captivity in al-Karak as follows: 170 infantry, 120 cavalry, 2 howitzers with their artillerymen, 150 mounted irregulars. This force would have been followed by a regiment of 500 men, had the operation not been called off.
27. Shuywayhat, *Al-'Uzayzat*, p. 76.
28. Tristram, *Moab*, p. 228.
29. Tristram, with his English background, used the word 'king' when he wrote about a visit to his camp by Fandi al-Fayiz. *Moab*, p. 227.
30. 'Badw', *Encyclopedia of Islam*, vol. I, p. 875.
31. The Salname Vilayet-i Suriya for the year 1308 H/1890 AD mentions Qada' al-Salt as having only six villages and one nahiya, Zizia, which has 19 villages.
32. Conder,*Survey*, gives the grand total of Bani Sakhr in the Balqa' as 1,505 tents or 7,500 people. The Bani Sakhr of the north were of equal number by general consensus.
33. Ibid., p. 295.
34. Musil, *Hegaz*, p. 36.
35. Al-Zirikli, *'Aman*, p. 60.
36. Napier, *War in Syria*, p. 185. Colonel Alderson, an eye-witness of the arrival of Ibrahim's troops at Gaza, estimated that round 8,800

regular troops were missing in addition to a roughly similar number of irregular troops, among whom there were some Egyptians.

37. Tristram, *Moab*, p. 320.
38. Al-'Arif, *Bi'r al-Sabi'*, p. 193.
39. This incident was related by Shaykh 'Awad al-Sattam at Sahab on 8 August 1985; he said that he heard it from his older brothers. He could not remember exactly when it happened but thought it took place some 15 years or so before his birth in 1891.
40. Mules were normally produced by a male donkey and a mare. The process was always disliked by beduins who therefore had no mules to speak of. Since female mules are sterile, farmers employed them in farming in exactly the same manner as male mules.
41. Technically a kadish is a male horse whose breed is not known to be of good origin. Normally it was not used for reproduction purposes with Arab mares. A kadisha is a female mare likewise of doubtful origin but the term was specifically used for any mare that bore a mule from a donkey. The whole group of kudsh was used in agriculture like mules.
42. People of al-Salt were already farming in al-Yaduda and also in Amman. This is confirmed by Merrill, who visited it on 27 April 1876: 'The Temple is occupied at present, by peasant families from al-Salt who are cultivating land in this vicinity,' *Jordan*, p. 264.
43. This information, in general, was kindly provided by Shaykh Ghalib al-Nawwaf at al-Qastal on 1 September 1985.
44. A very interesting remark about the fallow system is provided by Tristram, *Moab*, p. 121: 'The country was all a level rolling plain, very heavy after the rain, ploughed and sown in patches here and there, the rest sprinkled with herbaceous plants, and tufts of grass and stones, much as a neglected fallow might be at home, for the Arabs take one crop and leave the spot fallow for three or four years, while they scratch up the next patch.' These were definitely the days when there were fields that the population could till freely.
45. Zizia, a Hajj station 14 hours south of Zarqa, was mentioned as Qal'a al-Balqa' (al-Balqa Castle) in report no. 18, dated 27 Safar 1251 H/25 June 1835, presented to Muhammad 'Ali Pasha of Egypt after a survey was made by officers of the Egyptian Army of the Hajj route between Damascus and Makka. Two large water-pools were mentioned and it was reported that they both needed cleaning and repair. Al-Qastal (an Arabic word that might have originated from the Latin *castellum*) was a few kilometres to the west of the Hajj route; its large water-pool is still capable of retaining a huge volume of water in good rainfall years, exactly like the two at Zizia.
46. The per capita yearly consumption of wheat in this case has been considered on the basis of 125 kilograms. This is rather lower than other estimates, but it has been found necessary to take into consideration the fact that smaller-bodied people in the warm climates – when the men average 4 hours work a day – may require an average of 1,625 calories, while larger-bodied people in colder climates with men averaging 8 hours work a day may require 2,011 calories; Clark and Haswell, *Subsistence*

Agriculture, p. 58. They estimate annual per capita consumption at 190–250 kilograms. Hütteroth and Abdulfattah, *Historical Geography*, give 150–200 kilograms, while the *Survey of Palestine* for the Anglo-American Committee 9 December 1945, vol. II, gives around 200 kilograms.
47. Başbakanlik Arşiv Genel Müdürlügü, Yildiz Documents Kissim 93, Evrak 553/90, Karton 33.
48. A professor in the Iktisat Fakiltesi (economic faculty) at the University of Istanbul, who taught for two years at Yarmuk University in Irbid, Jordan.
49. Schumacher, *Jaulan*, pp. 86–93, gives a list of over 1,000 tent encampments headed by the amirs of 'Arab al-Fadil, as well as 300 tents of a Turkoman tribe.
50. A good account of the life of these Circassians in the Jawlan is given by Schumacher, op. cit.
51. Tristram, *Moab*, p. 344.
52. Rashid, *Malamih*, p. 84.
53. At present the speaker of the House of Deputies in the Jordanian parliament.
54. Please refer to chapter 11.

Chapter 11

1. Oliphant, *Gilead*, p. xiv.
2. Ibid., p. 51.
3. Ibid., p. 251.
4. This bureau was established in Istanbul to organize the settlement operations of refugees. The first mention of such a body is made in the 1297 H/1879 AD Salname-yi devlet-i aliyyeh, p. 91. In 1299 H it became a Muhajirin/Kumisiyun Ha'iti (Immigrants Commission Bureau) and in 1305 H it was transformed into an Idara 'Ummumiya (Directorate General).
5. Dagramache, quoting Ottoman documents, of Hacettepe University in Ankara, mentions these items of information in 'Immigration', p. 283.
6. Reports regarding the number on board vary between 2,000 and 3,000. The following summaries give some idea of the difficulty encountered in verifying data concerning events in those days, even when official reports of European governments are involved:

(a) The captain of the *Sphinx*, in his report, written and signed by Lieutenant E. G. Festing RN, commanding officer of HMS *Coquette*, mentions that the number was about 3,000. The report is at the Public Records Office, London, Admiralty and Secretariat papers (ADM 1/6445), undated but probably written on 9 March 1878.

(b) In his report to the secretary of the Admiralty no. 473, dated 18 March 1878, reporting proceedings of *Coquette* (Public Records Office ADM 1/7445), Vice Admiral G. Phipps Hornby, Commander in Chief, confirms that the telegram from H.B.M. consul at Larnaca to H.B.M.

consul-general at Beyrout stated that an Austrian Lloyd's steamer with 2,000 refugees on board had gone ashore on the Island of Cyprus.

(c) The report of the disaster published in *The Times* on Monday 11 March 1878 reads as follows:

> *Wreck Trieste March 9*
> The Austrian Lloyd steamer *Sphinx* from Kavala with 2,500 Circassians on board caught fire and went ashore near Cape Elia. Five hundred persons perished and the remainder were rescued.

(d) Evidently the papers of the *Sphinx* (written *Sphynx* by the owners) were completely lost. Lloyd Triestino stated in a letter dated 5 March 1984 that unfortunately the report on the wreck no. 364, dated 6 April 1878, could not be traced. It is not therefore possible at this stage to know precisely how many people boarded the vessel. The exact number of those who perished will never be known.

7. The Naval Historical Library, Ministry of Defence, London, kindly gave the following information about the *Sphinx* on 16 December 1983: 'According to the *Repertoire Générale de la Marine Marchande*, published by the Bureau Veritas Brussels in 1877, the *Sphinx* was a 3-masted iron schooner of 1,152 tons gross, 765 tons net. She was built on the Clyde in 1869 and had 150 HP engines. Her dimensions were: length 249 feet 1 inch, beams 31 feet 1 inch and depth 20 feet 5 inches. She was owned by Lloyd Austriaco and was registered at Trieste.' The vessel must have been overcrowded even if only 2,000 people were on board.

8. The main report of this disaster, at present in the Guildhall Library, appeared in Lloyd's list for 4 April and was dated 19 March 1878, Larnaca. It was kindly supplied by the Shipping Information Services Group, Lloyd's Register of Shipping, on 7 October 1983.

9. The report of the captain of the *Sphinx* confirms that the French gunboat was the *Linois*.

10. The Naval Historical Library, Ministry of Defence, London, kindly gave the following information on 16 December 1983 about HMS *Coquette*: A composite screw gunboat 430 tons, length 125 feet, beam 23 feet, built at Pembroke Dock and launched on 5 April 1871.

11. The arrival of the Circassians in Transjordan is not well documented. Besides different dates, very little is mentioned in such books as Peake, *Tarikh* and al-Mufti, *Abatira*.

12. Conder, *Heth and Moab*, p. 162. The survey of Wadi al-Sir and its environs was carried out during October 1881 and there was no mention of a Circassian settlement although the Circassians in Amman were referred to.

13. These circles or roundabouts provide street crossings from the Jabal Amman westwards. In 1948 there were only the first and second circles and there are now eight. With time, the urban sections extended gradually towards Wadi al-Sir; this municipality has jurisdiction over the area starting at the sixth circle.

14. In addition to immigrants from these two tribes, Circassians in

Transjordan include members of three other tribes: Bzadugh, Abaza and Abzagh.

15. Al-Mufti, *Abatira*, p. 229.
16. George Adam Smith visited them in 1891; *Historical Geography*, p. 20.
17. Al-Mufti, *Abatira*, p. 1905.
18. Peake, *Tarikh*, p. 370.
19. It is important to state that these two people are different and their languages are mutually unintelligible. The Circassians, who call themselves Adiqah, come from the western slopes of the Caucasus on the coast of the Black Sea. The Chichen come from Daghestan at the eastern end of the Caucasus range.
20. Weightman, 'The Circassians', p. 93.
21. Seteny, 'Anthropological research', p. 4.
22. Brawer, 'Circassian settlements', p. 2. The only religious minority in Transjordan were the Christians, who belonged all through the nineteenth century to the settled farmer group (who only took their flocks to the badiya in spring) and could not therefore be described as rebellious.
23. Luke and Keith-Roach, *Handbook*, p. 403.
24. Smith, *Geography*, p. 93.
25. Al-Mufti, *Abatira*, p. 229; Haghanduka, *Al-Sharkas*, p. 42.
26. It was a reasonably sized hardware store where most of the goods were bought wholesale in Jaffa and Jerusalem and brought by truck to Amman.
27. Actually there was a certain degree of disdain among the Circassians for trade and traders. This they shared with the nomadic tribes, although they held very different views regarding agricultural activity, which the Circassians respected and the beduins, in general, despised.
28. Tristram, *Moab*, p. 306.
29. Ibid., p. 307; Merrill, *East of the Jordan*, pp. 252, 472, 475.
30. Like the other tribes, the Bani Hassan were already deeply involved in agriculture by 1879. 'Starting with 'Ain al-Ghazzal (seven kilometres north of Amman) the stream becomes almost immediately large enough to be used for irrigating purposes, and the Arab cultivation begins from this point and continues down to the Jabbok, where that stream is used in like manner and irrigates the level bed of the valley', Oliphant, *Gilead*, p. 236. The Bani Hassan were and are still considered one of the largest Transjordanian tribes. Peake, *Tarikh*, p. 333, states that 'they have been in continuous struggle against the Bani Sakhr throughout the nineteenth century'.
31. Peake, *Tarikh*, p. 257, states that their grandfathers came to al-Balqa' from Syria around 1675 and as such are one of the oldest Transjordanian tribes. Four other tribes joined them to form al-Balqawiya alliance. Conder, *Survey*, p. 293, mentions that their principal shaykh was Sayil Ibn Haddid and that they had black slaves called *duraywat*.

32. Rashid, *Malamih*, p.84.
33. Members of al-Balqa' confederation included al-'Adwan, the people of al-Salt, al-'Ajarma, al-Balqawiya al-Da'ja, Bani Hassan, and the Christian tribesmen of Madaba. The allies of Bani Sakhr were al-Sirhan, Bani Hamayida, al-Hajjaya and 'Abad.
34. Musil, *Arabia*, English version, p. 5, quoted Nuri ibn Sha'alan shaykh mashayikh al-Ruwala declaring on 17 November 1908, 'For years I have been feeding my stock upon the lands of Ahl as-semal [Bani Sakhr]. During the harvest I camped at Libben [five kilometres from al-Hadid's domain and twelve kilometres from Amman] and we and our animals ate all that we found upon their lands and in their villages for we had no fear of anybody.' Although possibly true, with regard to the physical presence of al-Ruwala in al-Balqa', the statement contains a good deal of bravado and vainglory. The many battles between the two tribes in which the Bani Sakhr fared well, and the fact that al-Ruwala could not maintain any presence in Transjordan, then, give a clearer picture of the general situation.
35. Oliphant, *Gilead*, p. 206.
36. Libbey and Hoskins, *The Jordan Valley*, vol. I, p. 215.
37. Ibid., p. 216.
38. From a chapter by Guy Le Strange entitled 'A ride through Ajlun and the Belka', in Schumacher, *Across the Jordan*, p. 316.
39. The translation was kindly checked by His Lordship Fayiq Halazun, retired judge from the Court of Appeal in Amman.
40. The document was written and witnessed by Shaykh Mustafa Zayd al-Qadiri al-Kaylani, judge of al-Salt and witnessed by the following dignitaries: Munib 'Abd al-Razzaq Tukan, a dignitary of Nablus who transferred his residence to al-Salt; Iskandar Kassab, a Lebanese who was a high official in the fiscal section of the Ottoman administration in Palestine and had money-lending activities in al-Salt; Ahmad Abu Nuwwar, a Salti dignitary of the 'Atiyat clan and grandfather of General 'Ali Abu Nuwwar, commander of the Jordan Army 1956–7; Da'ud 'Asfur, the grandfather of the prominent 'Asfurs of Amman, a leading business family; Rashid Duzak, a Turk who probably was an official of the Ottoman administration at al-Salt; Fayadh al-Husayn, son of a leading dignitary of al-Qutayshat clan in the town; Amin al-Nabulsi, of the prominent business family who originally came from Nablus. Their activities included purchase of land, farming and money-lending; Yusuf Ibrahim Muhyar, another dignitary from Nablus who migrated to al-Salt (his grandson, General Hikmat Muhyar, was commander of security forces in Jordan during the early 1970s); Musa al-Sha'aban, a dignitary of al-Salt whose activity was in real estate and money-lending; Husayn al-Subuh, probably then the leading dignitary of al-Salt, being the shaykh of al-Qutayshat clans and their allies, al-Hara.

Strangely, neither the sellers nor the buyers signed the document, but two witnesses, Sayyah al-Haddid and Salim al-Hananda, testified to the authenticity of the power of attorney for Nuwayran ibn Sayil al-Haddid to represent his father, the paramount shaykh who seems not to have been present in the council of sale.

41. Neither Sa'id nor Sa'd, sons of Farhan Abujaber, was born when the sale deed was promulgated, but they vaguely remembered the story being spoken about in family circles. They did not, however, recollect hearing that any money was actually paid by their grandfather. Mr 'Ilmi al-Nabulsi, a grandnephew of Amin al-Nabulsi, asserted his belief that no money was paid, because the whole deal was simply an attempt to save the lands of Amman from being confiscated by the Ottoman authorities and then allotted to the Circassians.

42. Waleed Tash, 'The Circassians'.

43. Before his retirement Mr Tash was ambassador in Canada, worked at the UN in New York and was director-general of the Ministry for Foreign Affairs in Amman.

44. Weightman, 'The Circassians', p. 93. The Ottomans took advantage of the new settlers, using them as militia: '300 Circassian horsemen served with the Turkish garrison in al-Karak after it was subdued in 1894'; Mufti, *Abatira*, p. 229.

45. Weightman, 'The Circassians', p. 93.

46. An elderly lady, Nur al-Bisharat, born in al-Salt in 1882, related on 11 August 1972 that when she was about 15 she saw a group of beduin riders parade on the outskirts of the town with the heads of two Circassians they had killed hung high on two of their spears. Doubtless such happenings occurred on both sides, but the numbers involved could not have been large by any standard.

47. It is to be noted that nearly 30 years had to pass before any marriages started between members of the two communities. Generally Circassian women, famous for their beauty, married Arab Muslim men. Intermarriage between Christians on the one side and Muslim Arabs and Circassians on the other simply did not take place, because of religious differences and traditions.

48. Mufti, *Abatira*, p. 275. There is disagreement among Circassian authors about the date of this battle. Tash, 'The Circassians', confirms that it took place in 1904; while Haghanduka, *al-Sharkas*, p. 46, maintains that it happened in 1910. Considering the conditions that prevailed in Transjordan after the start of the twentieth century, one is inclined to agree with Mufti's date. Furthermore, having been born around 1900, Mufti was in a better position to hear about these events while they were still fresh in people's memories.

49. Al-Mufti, *Abatira*, p. 275; Haghanduka, *al-Sharkas*, p. 48.

50. Please refer to chapter 6 on transport and marketing.

51. Haghanduka, *al-Sharkas*, pp. 39, 60.

52. Ibid., p. 72.

53. Smith, *Historical Geography*, p. 669; al-Mufti, *Abatira*, p. 275.

54. Haghanduka, *al-Sharkas*, p. 70.

55. Butler, Norris and Stoever, *Syria*, p. 6.

56. 'Umar Musa Tihbsum, a Qabartai Circassian, was born in Amman in 1903. He was a farmer until 1942, when as a result of the great demand for wheat by the British forces command he started buying

cereals and exporting them to Palestine.

57. Mirza Pasha was the recognized leader of the Circassians and the Chichen in Transjordan until his death on 24 April 1932, when he was succeeded by Sa'id Pasha al-Mufti, many times speaker of the Senate and prime minister. The date of Mirza's birth is not known; Reuter's mentioned in his obituary that he was 95 years old, so he could have been born around 1840. It is said that he was an outlaw and became renowned for his courage and bravery when in 1874 the Ottoman government enlisted him and his men for the war in Serbia. In 1877, with 2,000 men, he defended with 'Uthman Pasha the Bulgarian city of Plevna, and then withdrew after the surrender. Afterwards he joined the police force in wilaya Suriya and the Salname of 1312 H/1894 AD, p. 100, mentions a Bikbashi Mirza Bey as commander of the regiment of cavalry and in 1313 H he is mentioned as commander of police in Amman. Four years later copies of cables in the archives of the Department of Libraries, Documentation and National Archives in Amman mention him as general, commanding police forces in Beirut. During the first world war he fought gallantly in the attack against the Suez Canal and later mobilized a cavalry regiment of 1,200 Circassians that guarded the Hijaz railway against Arab attacks after the start of the Arab revolt in 1916. When Amir 'Abdullah came to Jordan in 1921, Mirza Pasha became one of his ardent supporters and the Circassian community has served the new regime faithfully ever since. There seems to be general agreement that the smooth operation of land distribution in Transjordan was largely due to Mirza's wisdom and the fact that as a leader he was respected and accepted.

58. There is no record as to how this system was effected, but elders mentioned that newly-weds or families with one child were given 60 dunums. Families with two or three children were given 90 dunums, while bigger families were given the largest share of 120 dunums. In the last category, consideration was given for the maintenance of old parents and relatives and the presence of handicapped persons in the family.

59. Wadi Saqra is a fashionable quarter in west-central Amman. Until 1950 it was composed of rich fields which extended towards the area cultivated by the Circassians of Wadi al-Sir.

60. 'Ghiwar' was probably used to denote a Christian from al-Salt who possibly owned a field in the area.

61. It is interesting to note that an earlier owner was not mentioned since the land was given by the state without consideration to the claims made by third parties.

62. As no money was paid for the land, the only return that was expected by the Ottoman government was the regular payment of the yearly tithe. This is yet another proof of the great importance the state attached not only to the collection of taxes but also to the development of agricultural production so that more taxes could be collected.

63. Mr Sawbar, at present the mayor of Wadi al-Sir, is nearly 68 years old and kindly gave an interview on 15 April 1985.

64. Mr Khamash is a dignitary of Na'ur and is nearly 65 years old. He

kindly gave an interview on 11 April 1985. As a farmer, he still cultivates his fields in Na'ur and those of the Abujabers in al-Yaduda on a partnership basis.
65. These figures, given for the Circassians, on 10 March 1986 by Mr 'Umar Kalimat, chairman of the Circassian Philanthropic Society in Amman, and for the Chichen on 3 March 1986 by Mr 'Abd al-Ghani Hasan Husni, seem to be generally accepted. Since there is no official census, the best means of verifying them was to check the estimates with other dignitaries. The result was positive and it is therefore believed that these figures, with a margin of 10 per cent, can be considered as most probably correct.

Chapter 12

1. Harding, *The Antiquities of Jordan*, p. 73. Mr Harding served for 20 years with the Department of Antiquities and was between 1946 and 1956 its director-general.
2. Walsh, *The Moabite Stone*, p. 21.
3. Hoade, *East of the Jordan*, p. 98.
4. Seetzen, *Brief Account*, p. 37; Burckhardt, *Travels*, p. 365.
5. Tristram, *Moab*, p. 307.
6. Al-Shuwayhat, *Al-'Uzayzat*, p. 19.
7. Ibid; refers to a manuscript entitled 'Al-khatarat al-muqayyada', by Anastas-Marie al-Karmili.
8. Shahid, *Byzantium*, pp. 289 and 293.
9. Peake, *Tarikh*, p. 88.
10. Al-Shuwayhat, *Al-'Uzayzat*, p. 50. Many orthodox Christians joined the Roman Catholic, Latin and Protestant churches after the 1850s because of the activity manifested by these churches and the widespread disenchantment with the negligence and backwardness of the Orthodox patriarchate of Jerusalem and its Greek higher clergy. Only a few changed course after a while and returned to the Orthodox church. There is only one case on record of a Latin leaving his religious denomination and joining the Orthodox church. This was Salih Abujaber in 1878 who had a fight for influence with the Latin priest in al-Salt and the Latin patriarch in Jerusalem. His own brother did not join him.

A good description of the state of affairs in the two churches is given by Tristram, *Israel*, p. 266: 'There is all the difference between the monks of the Greek and Roman rites in Palestine that characterizes the political and religious position of the two churches. The one is always on the aggressive, the other on the defensive. In everything Greek there seems embodied a cold, dead conservatism, tenacious it knows not why, and obstinate looking on every concession or relaxation of a rule as a confession of weakness.'
11. The Orthodox patriarch from 1875 to 1882 was Erothios. By 1880 he was under pressure to give more privileges to the Arab Orthodox in

the running of the affairs of the patriarchate. Khuri, *Khulasat*, p. 2220.

12. The Latin patriarch of Jerusalem 1873 to 1889 was Monsignor Vincenzo Bracco, known to the Arabs in Palestine and Transjordan as Batrak Mansur; Possetto, *Il Patriarcato*, p. 231.

13. Father Alexander was an Italian, Don Alessandro Maccagno. He joined the Christian tribesmen in their migration to Madaba in 1880; ibid., pp. 287 and 307. Father Bulus, also an Italian, was Don Paolo Bandoli. He assisted Don Alessandro in al-Karak and Madaba for two years. Medebielle, 'La difficile installation', p. 23.

14. Al-Shuwayhat, *Al-'Uzayzat*, p. 50.

15. Saba and 'Uzayzi, *Madaba*, p. 144.

16. During the latter part of the nineteenth century the divisions given in the table were listed by al-Shuwayhat regarding allegiance (*Al-'Uzayzat*, p. 43), and by Jaussen regarding numbers of khanas for each tribe (*Coutumes*, p. 394), as well as by Saba and al-'Uzayzi (*Madaba*, p. 140): As far as land distribution was concerned, the Christians as a group were allocated one-third, while the Muslims of both alliances received one-third each. Al-Qusus, 'Mudhakkirat', p. 5.

Khana	Tribe	Religion
Al-Gharaba		
140	al-Majali	Muslim
240	al-Ma'ayita	Muslim
130	al-Habashna	Muslim
50	al-'Amr	Mulsim
40	al-'Uzayzat	Christian
30	al-Ma'aya'a	Christian
20	al-Zuraykat	Christian
Al-Sharaqa		
200	al-Tarawna	Muslim
160	al-Dumur	Muslim
160	al-Sarayra	Muslim
40	al-Halasa	Christian
30	al-Haddadin	Christian
20	al-Hijazin	Christian
15	al-'Akasha	Christian
30	al-Karadisha	Christian
10	al-Madanat	Christian

17. Burckhardt, *Travels*, p. 382, describes Christian valour and how 27 of them defeated over 400 of the Ruwala in the year 1812. Mention is also made of the men of al-Karak as generally being excellent marksmen. During the first four decades of the twentieth century the Christian tribesmen of Madaba also made a good name for themselves in

fights with the Bani Sakhr and the Balqa' clans, both of whom were much larger in numbers. Their shaykh, Hanna Ibn Farah, was famous for his courage and it was generally accepted that 'he never missed a shot in his life'.

18. He was an able Ottoman administrator who served as maktubji (secretary) Wilaya Suriya in 1888 and was appointed as mutasarrif of al-Karak on 3 Mart 1309 H/3 May 1892 in accordance with the Salname Vilayet-i Suriya for 1317 H (p. 229). Sadiq Pasha, his successor to the post, was appointed on 13 Aylul 1312 H/13 August 1894 AD.

19. Saba and al-'Uzayzi, *Madaba*, p. 145.

20. Midhat Pasha, the leader of the Ottoman liberals, was relieved from the grand vizirate by Sultan 'Abd al-Hamid in 1877 when the Ottoman Chamber was dissolved. He was appointed as wali of Damascus on 12 Tishrin Thani 1294 H/12 November 1877 AD and remained in this post for 1 year, 8 months and 10 days. Salname Vilayet-i Suriya 1317 H.

21. Al-Shuwayhat, *Al-'Uzayzat*, p. 68.

22. Nablus was at this time (1880) the administrative centre of sanjaq al-Balqa' including al-Salt and Madaba.

23. Saba and al-'Uzayzi, *Madaba*, p. 148.

24. This was the department where fiscal records of revenue and expenditure were kept. Prior to the Tanzimat period in 1839 it was called the Daftar Khana. Evidently the records in Damascus were checked and when it was ascertained that there were no registered owners for Madaba, the permission was granted for its settlement by the three tribes.

25. The originals of these letters are either in the archives of sanjaq Nablus or the Latin patriarchate of Jerusalem. Since both Jerusalem and Nablus are now under Israeli occupation and a visit to neither could be arranged, it was only possible to refer to the texts of the letters in Saba and 'Uzayzi, *Madaba*, pp. 144–9.

26. Al-Shuwayhat, *Al-'Uzayzat*, p. 75.

27. Estimating that Madaba had 100 faddans, the khawa to be received by Sattam could have been – considering the camel load to be 48 sa's of nearly 6 kilograms each – as follows:

4,800 sa's x 6.0 kg =	28,800 kg	wheat
4,800 sa's x 5.5 kg =	26,400 kg	barley
4,800 sa's x 6.0 kg =	28,800 kg	sorghum
Total	74,000 kg	cereals

This would have been nearly a 15% addition to the crop he produced on his own estate.

28. Al-Shuwayhat, *Al-'Uzayzat*, p. 77.

29. Al-Hummar became a summer resort for the late King 'Abdullah and during the 1960s King Husayn built two palaces there.

30. A family of important merchants in Nablus who still own one of the largest soap factories in the town. One of them, Bassam al-Shak'a, was until recently the mayor of Nablus.

31. Al-Shuwayhat, *Al-'Uzayzat*, p. 77.
32. Fights took place every now and then, especially during the difficult years of the first world war. Ishaq Ibn Farah was a casualty of one of these skirmishes in which a few men of both sides were killed and he became lame for life.
33. Robinson Lees, *Jordan*, p. 60.
34. True to beduin traditions, the Christians of Madaba strengthened their numbers through marriage of their daughters to strangers who were then inducted into the tribes. Information kindly supplied by Ruks al-'Uzayzi reveals that at least four such clans developed in this manner: Ibn Farah married into al-Sawalha; Al-Masarwa married into al-Shuwayhat; Al-Masri married into al-Zawayda; Tannus married into al-Tiwal.
35. The composition of the town's population has undergone important changes since the 1960s with people from the countryside and ex-Palestinian refugees becoming the majority of a population of nearly 35,000. The original tribes are estimated at around 20,000 but many of them live in Amman and abroad. At present the municipal council does not have a single member from them as a result of their boycott of elections.
36. The personal touch meant a continuous beaming smile, attention to the comfort of the guest, and better cooking since the settlers had better facilities than the nomads. In Arabic it is called *al-karam wa-al-tayb*. Jaussen, *Coutumes*, p. 129.
37. The usual practice in the encampments was to send a rider to bring a lamb from the flocks in pasture when a guest arrived. To avoid this delay, which might be two or three hours, the competitors kept a few lambs near the tents and killed one or more of them immediately the guests arrived.
38. The area subsequently increased; towards 1930 the tribesmen of Madaba were probably cultivating over 50,000 dunums, of which half was owned and the other half run on a partnership basis.
39. Information kindly given by Ruks ibn Zayid al-'Uzayzi who was nearly 80.
40. It is also worth noting that, in spite of the fact that they were mutually friendly Christian communities, no intermarriage happened until 1960 when Ghalib Salih Frayh Abujaber married Samiya Ishaq al-Farah. This was confirmed to me by my father, Sa'd Farhan Salih Abujaber during 1964.
41. Records of the Ministry of Agriculture in Amman state that the average was 250,000 sheep and 300,000 goats towards the end of the 1920s. However, as these figures are derived from tax statistics, it could well be that the actual sheep and goat population was double this as owners did not declare the correct figures. This belief is strengthened by the fact that more reliable figures for 1975, as given by the Department of Statistics in Amman, were 569,479 sheep and 326,309 goats. Since 1960 goat-keeping has been discouraged in an attempt to save small trees from their voracious appetites.

42. The same principles used in chapter 10 were applied to the computation.
43. Based on the computation for Abujaber flocks at the start of the twentieth century.
44. Arab mares were in great demand until motor cars became popular towards 1925. Their prices were high and a good one could fetch as much as £500. The average, however, was between £100 and £200. Stallions were not popular for riding and were generally kept for stud purposes.
45. A copy of a strongly worded petition submitted by Farhan Abujaber and his brothers against the unjust assessment of the 1915 taxes on al-Yaduda was found in the family papers in 1964. It claimed that al-Multazim Hanna ibn Farah and his partners in the iltizam were trying to double their earnings in one year.
46. Saba and al-'Uzayzi, *Madaba*, p. 255.

Chapter 13

1. Rustum, *Ara'*, p. 170.
2. Zaki, 'Hamla al-sham', p. 398. Peake, *Tarikh*, p. 180, mentions the Egyptian origin of many families in Transjordan besides those whose grandfathers were soldiers in the army of Ibrahim Pasha and stayed behind after its withdrawal, for example one clan which lived in al-Karak district.
3. There were sometimes two or three families who were called Masri or Masarwa and were not relatives; cf. al-Shuwayhat, *Al-'Uzayzat*, pp. 179 and 191, where mention is made of such families in Madaba. The same applies to Nablus in Palestine where two families carry the name without being related.
4. This story was told by Shaykh Munwir, grandson of Shaykh 'Abd Allah Abu Zayd. Born in 1926, he lives in Sahab and kindly related what he heard about the early days of Sahab. The interview was held on 14 November 1986.
5. Al-Dabbagh, *Biladuna Filastin*, part I, section II, p. 275. This book also confirmed the area of Hawj and that its population in 1922 during the British mandate in Palestine numbered only 426 people.
6. During interviews held at Sahab on 14 November 1986, this and other items of information were kindly passed on by Shaykh Ahmad Salama Abu Zayd who was born in 1910 and lives in Sahab. His son Yasin, who works with the Jordan Television service, arranged the meetings with the dignitaries in Sahab most pleasantly as well as helping with verification of dates and genealogies.
7. Tristram, *Moab*, p. 95, mentions that Sattam visited Cairo and Alexandria and used the railway during these visits.
8. There must have been other Egyptian families in Transjordan. Shaykh Munwir mentioned a big clan in al-Ramtha, and another at Umm Qays. His cousin Na'if Ibn Muhammad, whose grandfather

Ahmad was a younger brother of Shaykh ‘Abd Allah, confirmed that the ‘Abd al-Jawad and Abu Haswa clans were at al-Yaduda and Umm al-Basatin, al-Maharma at Jalul, and his own father at al-Yaduda where he was born in the early 1920s. The interviews were held in Sahab on 14 November 1986 and 6 June 1986 respectively.

9. During an interview on 17 July 1985 in Amman, Shaykh ‘Akif Ibn Mithqal al-Fayiz confirmed that his grandfather Sattam and his sons had as partners the clan al-Mara‘iba of Egyptian origin in Zizia. This must have been during the last two decades of the nineteenth century after Zizia was taken over as *jiftlik* or imperial holding by Sultan ‘Abd al-Hamid. During the same interview ‘Akif Bey confirmed that in 1921 half Zizia was owned by Amir ‘Abd Allah, one-quarter by Shaykh Mithqal, and one-quarter by Sa‘d al-Din Pasha Shatilla, a Lebanese. Amir ‘Abd Allah gave his half to Mithqal, who later went to Beirut and bought the last quarter from Shatilla.

10. These numbers were important at the time; Egyptians could have formed up to 25% of the farm-hands in the villages of Bani Sakhr and 10% at al-Yaduda which was the largest agricultural venture in the area.

11. According to information received from Mahmud Ibrahim al-‘Armiti at Sahab on 6 June 1986. Mr al-‘Armiti was born at al-Yaduda and was at least 65 years old. More information was received from Ahmad Salama Abu Zayd on 14 November 1986 at Sahab.

12. Bell, *The Desert and the Sown*, p. 35. Travelling in March/April 1899 she wrote: ‘The fort at Zizia which was repaired by Shaikh Sattam of the Sukhur has now fallen to the Sultan since it stands in the territory selected by him for his chiftlik.’

13. Related by Na’if Abu Zayd in Sahab on 6 June 1986.

14. Munwir Abu Zayd was not sure whether it was the qa’immaqam at al-Salt or the wali in Damascus. Yasin Abu Zayd said he always heard it told that it was the wali in Damascus.

15. Burckhardt, *Travels*, p. 370. He wrote in 1812 that they were beduins of the Arabian desert who resorted to al-Balqa’ in summer for pasturage. He also confirmed that they were a tribe of more than 5,000 tents.

16. Ibrahim, ‘Sahab’, p. 23.

17. These castles were built in the 6th, 7th and 8th centuries. The largest was built at Azraq, the only permanent body of water in 12,000 square miles of desert. Four of them, at Mushatta, Tuba, al-Kharana and ‘Amra, are examples of Islamic art and architecture. Generally they were used by the Ummayad caliphs as hunting lodges and pleasure palaces where life embodied the luxury of life in the palaces of Damascus. Only one among them, the castle at al-Kharana, was built for defensive purposes and that is evidently due to the fact that it was situated at the crossroads of ancient desert routes.

18. Al-Qutb, *Ansab al-‘Arab*, p. 59.

19. Ibid.

20. Related by Mr Muhammad Salim al-Fawri, originally from al-Salt but then working in the Land Registration Department in Amman. He

confirmed having heard it years ago from shaykhs of the Bani Sakhr.
21. Peake, *Tarikh*, p. 252.
22. Peake mentions in this and other chapters of his book that Ottoman administration was established in al-Karak in 1892, whereas it actually occurred in 1894. Official records of the Jordan government give the latter date.
23. 'Uthman Nuri Pasha ruled Damascus twice: the first time for 6 months and 21 days in 1889; the second for 2 years 11 months and 17 days, from July 1892 to July 1895. cf. Ibn Jum'a, *Wula Dimashq*.
24. Ahmad Salama Abu Zayd confirmed on 14 November 1986 that he had heard from his father that Shaykh 'Abd Allah conceded 1,000 dunums to Sa'd al-Ghrayr of al-Da'ja tribe and another 1,000 dunums to al-Raqqad clan of al-Balqa'.
25. Shaykh 'Abd Allah was known for his determination to have the largest possible number of Egyptian clans in Sahab and therefore his gesture to al-Maharma must be viewed from this angle.
26. The population of Sahab at the beginning of the twentieth century could not have been more than 80 families, or 400 people; this figure grew to 1,200 in 1940 and to around 3,000 in 1960 when the Municipal Council was elected for the first time. The town has now a population of nearly 15,000, of whom 12,000 are the descendants of the three original Egyptian clans.
27. Al-Muwaqqar, lying to the east of Sahab, has an area of 52,000 dunums (5,200 hectares, of which 4,700 are arable land) and is the property of al-Khurshan of Bani Sakhr. In 1921 their Shaykh Haditha succeeded in obtaining an *irada* (princely order) from Prince 'Abd Allah of Transjordan to have al-Muwaqqar's lands distributed to the members of the tribe. On 14 October 1921, the area was divided during a council (majlis) into 14 shares, of which Shaykh Haditha was allotted two shares and al-Maharma of al-Muwaqqar one share, which meant that the latter were considered affiliates of the tribe. However, due to objections from different tribesmen, another majlis was held on 25 June 1922 which decided that, since al-Maharma were Egyptians of Sahab, their shares should be withdrawn and reallocated to Bani Sakhr tribesmen. The beduin tradition that a cousin, however distant, is closer than a very good friend was once more decisive. The story is registered in detail in the *Daftar Aradi al-Muwaqqar* (Land register), now preserved at the Lands and Surveys Department at Amman.
28. Mr Na'if Abu Zayd confirmed on 6 June 1986 in Sahab that his own crop in one year was as follows:

60 sa's wheat seeds, crop 1,200 sa's (7,200 kg)
60 sa's barley seeds, crop 2,200 sa's (12,000 kg)

29. Shaykh Munwir Abu Zayd confirmed on 14 November 1986 at Sahab that he remembered that when he was a child every year a tax collector spent a week in the village and pleasantly agreed to the figure of 22,000 head of sheep and goats declared by the villagers, plus some

few hundred cows and mules, horses, oxen, and donkeys that constituted the village's animal work force.

30. Sahab has a much stronger involvement in trade nowadays since its location on the roads leading to Saudi Arabia and Iraq has given it quite an advantage over Madaba.

Bibliography

A Interviews

Besides information gathered from family members, interviews were conducted with the following: Hajj Tawfiq al-Riyalat, 70-year-old member of a Salti clan, 4 April 1981; Na'ur Ibn Falih Ibn al-Sayyid Ibn Rumayh Abu Junayb, 1 August, 1983; Sulayman Qamuh, member of a Salti clan, high official at Department of Lands and Survey, 3 August 1984; Norman Lewis, 25 October 1984; Muhammad al-Rashid, 30 October 1984 and 13 January 1985; His Lordship Fawwaz al-Rusan, land settlement Judge for nearly 40 years, 3 April 1985; 'Ali Mansur Wanas al-Hindawi, great-grandson of Hindawi, Amman, 6 April 1985 and 29 May 1986; Shaykh Mansur Wanas al-Hindawi, al-Nu'ayma, 9 April 1985; Sa'd al-Na'im al-Murayyan, al-Nu'ayma, 9 April 1985; Ishaq Yusuf Khamash, of Na'ur, 11 April 1985; Na'im Abu Sha'ar of the Numura clan, 13 April 1985; Husni Shuman Ahmad Sawbar, born 1919, mayor of Wadi al-Sir, 15 April 1985; Dr Fawzi Zayadin, archaeologist at the Department of Antiquities, Amman, April 1985; 'Umar Tihbsum, a Qabartai Circassian born in Amman 1903, 26 June 1985; Mujhim Ibn Haditha al-Khuraysha, governor of al-Salt, 6 July 1985; Shaykh 'Awad al-Fayiz, grandson of Fandi, 1 August 1985; Shaykh Ghalib Ibn Nawwaf, al-Qastal, 8 August 1985 and 1 September 1985; Shaykh Nawash Ibn Mithqal, al-Qastal, 1 September 1985; Ba'thi al-Jaradat of Kufrjayiz, 25 September 1985; Dr Da'ud al-Husayni, former Deputy of Jerusalem, Amman, 10 November 1985; 'Abd al-Ghani Hasan Husni, a Chichen, al-Zarqa', 3 March 1986; Mahmud Ibrahim al-'Armiti, Sahab, 6 June 1986; Na'if Abu Zayd, Sahab, 6 June 1986; Ahmad Abu Zayd, Sahab, 14 November 1986; 'Umar Kalimat, chairman of the Circassian Philanthropic Society in Amman, 10 March 1986; 'Ilmi al-Nabulsi, a grand-nephew of Amin al-Nabulsi; Ruks Ibn Zayid al-'Uzayzi; Shaykh 'Akif, grandson of Sattam, speaker of the House of Deputies, Jordanian Parliament, and Rauhi al-Khatib, mayor of Jerusalem.

B Unpublished written sources

Arabic

1 The papers (including accounts books and correspondence) at al-Yaduda. The collection of Abujaber documents of transactions are mainly held here (there are a few at al-Salt). They cover activities in many fields and vary from a loan of six sa's of wheat to the purchase of 4,000 dunums of the best arable land in Transjordan. Some date back to 1880. They thus provide an excellent source of information about the country.
2 Jarida al-da'awa wa-al-sukuk al-shar'iya al-waqi'a fi mahkama al-salt al-shar'iya (begun Rabi 'Al-Akhir 1304 H/1886 AD).
3 Daftar aradi al-muwaqqar (Land register), Department of Lands and Survey, Amman.
4 Land Court File, Department of Lands and Survey, Amman.
5 Directive no. 16/3/20/2275 to Land Registration Directors in Jordan, 15 April 1953, Department of Lands and Survey, Amman.
6 Records of the Ministry of Agriculture, Amman.
7 Records of the Department of General Statistics, Amman.
8 Copies of cables in the Department of Libraries, Documentation and National Archives, Amman.
9 Records in the Latin church in Nazareth.

Turkish

Başbakanlik Arşivi Genel Mudurlogo, Yildiz.

English

1 1929 Report of the Legion submitted by commander Peake Pasha, Maktaba al-bahth al-'ilmi, Ministry of Agriculture, Amman.
2 Public Records Office, London, FO 95/927 Syria and Palestine 1869; FO 195/675; FO 78/1978; ADM 1/6445.

C Official publications

1 His Britannic Majesty's Government, *Colonial Report no. 20*, to the League of Nations on the administration of Palestine and Transjordan for the year 1925.
2 Jordan, Central Water Authority, Hydrology Division, *30 Year Average Rainfall in Jordan 1931–1960*, Technical Paper no. 34, part I, December 1984.
2 Jordan, Natural Resources Authority, Hydrology Division, *Rainfall in Jordan in the Water Year 1966–1967*, Technical Paper no. 32, June 1968.
4 Jordan, Department of Lands and Surveys, Map of Village Index, 1973.

5 Palestine, Government Printer, *Survey of Palestine* for the Anglo-American Committee of December 1945 and January 1946, 1946.
6 Jordan, Department of Lands and Surveys, List of Surveyed Sites, Annual Report for 1975.
7 Jordan, Department of General Statistics, Census, 10 November 1979.
8 Brussels, Bureau Veritas, Répertoire général de la marine marchande, 1877.
9 Lloyds List for 4 April 1878.
10 Turkey, Sâlnâme-yi devlet-i aliyyeh-yi osmaniye, 1266 H/1849, 1272 H/1854, 1273 H/1855, 1285 H/1868, 1290 H/1872, 1297 H/1879, 1299 H/1881, 1305 H/1887.
11 Turkey, Sâlnâme Sūriya vilayet-i . . . 1288 H/1870, 1289 H/1871, 1290 H/1872, 1304 H/1886, 1308 H/1890, 1317 H/1899.

D Published works in Arabic

ʻAbdallah, Ibn al-Husayn (King Abdulla) *Mudhakkirat al-malik ʻabd allah*, Jerusalem, 1945.
Abu Rumayla, Baraket, *Al-aʻshab fi al-urdunn*, Amman, 1981.
Abu Yusuf, al-Qadi, *Kitab al-kharaj*, Beirut, 1939.
ʻAwad, ʻAbd al-ʻAziz Muhammad, *Al-'Idara al-ʻuthmaniya fi wilaya suriya 1864–1914*, Cairo, 1961.
al-ʻArif, ʻArif, *Al-mufassal fi tarikh al-quds*, Jerusalem 1961.
al-ʻArif, ʻArif, *Bi'r al-sabiʻ wa-qaba' iluha*, Jerusalem, 1934.
al-ʻAwra, Ibrahim, *Tarikh wilaya sulayman basha al-ʻadil*, Saida, 1936.
Bakhit, M. A. et al. eds, *Kashf ihsa' li sijillat al-mahakim al-sharʻiya wa-al-awqaf al-islamiya fi bilad al-sham*, Amman, 1984.
Banura, Tuma, *Tarikh Bayt lahm, Bayt jala wa-Bayt sahur*, Jerusalem, 1982.
Bayhum, Muhammad Jamil, *Al-halaqa al-mafquda fi tarikh al-ʻarab*, Egypt, 1950.
Burayk, Mikha'il, *Tarikh al-sham*, ed. Qustantin al-Basha, Harisa (Lebanon), 1930.
Chatty, Dr Dawn, 'Tawasuʻ al-badw fi bilad al-sham wa inhisaruhum', in *al-Muʻtamar al-duwali al-thani li-tarikh bilad al-sham*, Damascus, 1979, vol. 1.
al-Dabbagh, Mustafa Murad, *Biladuna Filastin*, 11 vols, Dar al-taliʻa, Beirut, 1965–76; vol. 12, R.S.S. Press, Amman, 1986.
Farid, Muhammad Bey, *Tarikh al-dawla al-ʻaliya al-ʻuthmaniya*, Beirut, 1977.
Gharayba, ʻAbd al-Karim, *Qiyam al-dawla al-saʻudiya al-ʻarabiya*, Cairo, 1974.
Gharayba, ʻAbd al-Karim, *Suriya fi al-qarn al-tasiʻ ʻashar*, Cairo, 1962.
Ghawanma, Yusuf Darwish, *Sharq al-urdunn fi ʻasr dawla al-mamalik*, 2 vols, Amman, 1979.
Ghawanma, Yusuf Darwish, 'Al-taʻun wa-al-jafaf wa atharuhum fi al-ʻasr

al-mamluki', in *Studies in History and Archeology of Jordan*, Amman, 1985, vol. II, p. 315.

Haghanduka, Muhammad Khayr, *Al-sharkas, 'asluhum, tarikhuhum, hijratuhum ila al-urdunn*, Amman, 1982.

al-Hasani, 'Ali 'Abd al-'Aziz, *Tarikh suriya al-iqtisadi*, Damascus, 1923.

Hinz, Walther, *Al-makayil wa al-awzan al-islamiya*, trans. from German by Dr Kamil al-'Asali, Amman, 1980.

al-Husari, Sati', *Al-bilad al-'arabiya wa-al-dawla al-'uthmaniya*, Cairo, 1957.

Ibn Jum'a, *Wula Dimashq fi al-'ahd al-'uthmani*, ed. Salah, Damascus, 1949.

Ibn Tulun, Shams al-Din Muhammad, *I'lam al-wara*, Damascus, 1964.

Ibn Tulun, Shams al-Din Muhammad, *Mufakahat al-khilan fi hawadith al-zaman*, 2 vols, Cairo, 1964.

al-Jamil, Makki, *al-Badu wa-al-qaba'il al-rahhala fi al-'iraq*, Baghdad, 1956.

Khuri, Shihada and Nicola, *Khulasat tarikh kanisa urushalim al-urthuduksiya*, Jerusalem, 1925.

Kurd, 'Ali Muhammad, *Khitat al-sham*, 6 vols. 3rd edn, Damascus, 1983.

al-Madi, Munib and Musa, Sulayman, *Tarikh al-Urdunn fi al-qarn al-'ishrin*, Amman, 1959.

Makarius, Shahin, 'al-Ma'arif fi Suriyyah', *al-Muqtataf*, vol. 7 (1882/3) 385–92, 465–76, 529–37.

al-Manawi, Muhammad Ibn 'Abd al-Ra'uf, *Al-nuqud wa al-makayil wa-al-mawazin*, ed. Dr R. M. Samarrai, Baghdad, 1981.

Mansür, Asad, *Tarikh al-Nasira*, Cairo, 1924.

al-Maqrizi, Ahmad Ibn 'Ali, *Kitab al-suluk li-ma'rifa duwal al-muluk*, ed. Sa'id 'Ashur, 10 vols, Cairo, 1972.

al-Mu'ammar, Tawfiq, *Dahir al-'Umar*, Nazareth, 1979.

al-Mufti, Shawkat, *Abatira wa-abtal fi tarikh al-qalqas*, Jerusalem, 1962.

al-Muqaddasi, Shams al-Din Abu 'Abd Allah, *Ahsan al-taqasim fi ma'rifa al-aqalim*, Leiden, 1877.

Nawfal, Nawfal Ni'ma Allah, *Al-dustur*, Beirut, 1301 H.

Peake, Lt Col. F. G., *Tarikh sharq al-urdunn wa-qaba'iluha*, trans. of *History and Tribes of Jordan* by Baha al-Din Tuqan, Jerusalem, 1934. (The Arabic version of Peake's work has usually been cited in preference to the original English on the grounds that the translator corrected the whole work and also read all the Arabic sources referred to in the book, thus producing a more comprehensive edition. The translator still lives in Amman and was available for consultation.)

al-Qusus, 'Awda, 'Mudhakkirat 'Awda Qusus 1877–1943', unpublished MS, al-Karak, 1920s.

al-Qutb, Samir 'Abd al-Razzaq, *Ansab al-'Arab*, Beirut, 1969.

Rafiq, 'Abd al-Karim, *Bilad al-sham wa-misr 1516–1798*, 2nd edn, Damascus, 1968.

al-Ramini, Akram, *Nablus fi al-qarn al-tasi' 'ashar*, Amman, 1980.

Rashid, 'Abd Allah, *Malamih al-hay al-sha'biya fi madina amman 1878–1948*, Amman, 1983.
Rustum, Asad, *Al-Mahfuzat al-malikiya al-misriya*, 4 vols, Beirut, 1943.
Rustum, Asad, *Ara' wa-abhath 1897–1965*, Beirut, 1967.
Saba, Father George and al-'Uzayzi, Ruks Ibn Zayid, *Madaba wa-dawahuha*, Jerusalem, 1961.
Salman, Bishop Bulus, *Khamsa a'wam fi sharq al-urdunn*, Harisa (Lebanon), 1929.
Shahin, 'Aziz, *Kashf al-niqab 'an al-judud wa'al-ansab fi ramallah*, Jerusalem, 1982.
Shidyaq, Shaykh Tannus Ibn Yusuf, *Akhbar al-a'yan fi jabal lubnan, 1859*, Beirut, 1954.
al-Shuwayhat, Dr Yusuf Salim, *Al-'Uzayzat fi madaba*, Amman, 1964.
Sipano, Ahmad Ghassan, *Tarikh hawadith al-sham (al-dimashqi 1772–841)*, Damascus, 1982.
Sipano, Ahmad Ghassan ed., *Mudhakkirat tarikhiya 'an mamlaka ibrahim basha 'ala suriya*, 2 vols, Damascus, after 1981.
al-Sudayri, Muhammad Ibn Ahmad, *Abtal min al-sahra'*, Riyadh, 1983.
Taqi al-Din, Adib, *Muntakhabat tawarikh dimashq*, Damascus, 1927.
Tusun, 'Umar, *Al-jaysh al-misri al-barri wa-al-bahri*, Cairo, probably 1930.
al-'Uzayzi, Ruks Ibn Zayid, *Qamus al-'adat wa-al-lahjat wa-al-'awabid al-urdunniya*, Amman, 1973.
Wallin, George August, *Suwar min shamal jazira al-'arab*, Beirut, 1971.
Zaki, 'Abd al-Rahman, 'Hamla al-sham al-ula wa-al-thaniya' in Royal Society for Historical Studies (ed.), *Dhikra al-batal al-fatih ibrahim basha*, Cairo, 1948.
al-Zirikli, Khayr al-Din, *'Aman fi 'amman*, Cairo, 1925.

E Published works in European languages

Adam Smith, George, *The Historical Geography of the Holy Land*, 12th edn, London, 1906.
Bakhit, Muhammad Adnan, *The Ottoman Province of Damascus in the Sixteenth Century*, Beirut, 1982.
Bell, Gertrude, *The Desert and the Sown*, London, 1919.
Bale, G. S. and Goldschmidt, M. J., *Geology and Water Resources of Palestine*, Government of Palestine, Jerusalem, 1947.
Blake, G. S. and Goldschmidt, M. J., *Geology and Water Resources of Palestine*, Government Printer, Jerusalem, 1947.
Bowen, Harold, 'Akçe' in *Encyclopaedia of Islam*, new edn, vol. 1, Leiden, 1960.
Bowring, John, *Report on the Commercial Statistics of Syria*, London, 1840.
Brawer, Moshe, 'Circassian settlements in Galilee', unpublished paper presented to the Symposium on Minority Group Settlement, 22nd International Geographical Congress, Winnipeg, July, 1972.

Buckingham, J. S., *Travels among the Arab Tribes*, London, 1825.
Burckhardt, J. L., *Travels in Syria and thëHoly Land*, London, 1822.
Butler, Howard Crosby, Norris, Frederick A. and Stoever, Edward Royal, *Syria*, Publications of the Princeton University Expeditions to Syria in 1904–5 and 1909, Division Geography and Itinerary, Leiden, 1930.
Canaan, T., 'The Saqr Bedouin of Bisan', in *Journal of the Palestine Oriental Society* 16 (1936) 21.
Church Missionary Society, *Register of Missionaries from 1804 to 1904*, printed for private circulation, London, 1904.
Clark, Colin and Haswell, Margaret, *The Economics of Subsistence Agriculture*, London, 1970.
Cohen, Amnon and Lewis, Bernard, *Population and Revenue in the Towns of Palestine in the 16th Century*, Princeton, NJ, 1978.
Conder, Claude Reignier, *Heth and Moab*, London, 1883.
Conder, C. R., *The Survey of Eastern Palestine*, vol. I, *The Adwan Country*, The Committee of the Palestine Exploration Fund, London 1889.
Coon, Carleton S., 'Badw', in *Encyclopaedia of Islam*, new edn, vol. II, Leiden, 1960.
Dagramache, Dr Amal, 'Immigration to Syria at the end of the 19th century', paper submitted to the Second International Conference on the History of Bilad al-Sham, University of Damascus, 27 Nov.–3 Dec. 1978, Tarabish, Damascus, 1979.
Diab, Henry and Wahlin, Lars, 'The geography of education in Syria in 1882. With a translation of "Education in Syria" by Shahin Makarius, 1883', *Geografiska Annaler* 65B (1983) 105–27.
Dols, Michael, W., *The Black Death in the Middle East*, Princeton, NJ, 1977.
Doughty, Charles M., *Travels in Arabia Deserta*, 2 vols, New York, 1979.
Dudik, Mudia and Soda, *Firearms*, Prague, 1981.
El-Edroos, Brigadier Syed Ali, *The Hashemite Arab Army 1908–1978*, Fakenham, Norfolk, 1980.
El-Sherbini, A. A., *Food Security, Issues in the Arab Near East*, United Nations Economic Commission for Western Asia, 2 vols, Oxford, 1976.
Fantechi, R. and Margaris, N. S. eds, *Proceedings of the Information Symposium in the EEC Programme on Climatology held in Mytilene, Greece, 1984*, Dortrecht, 1986.
Faroghi, Suraiya, *Towns and Townsmen of Ottoman Anatolia*, Cambridge, 1984.
Farwell, Byron, *A Biography of Sir R. F. Burton*, London, 1963.
Finn, James, *Stirring Times*, 2 vols, London, 1878.
Finn, Mrs James, 'The fallaheen of Palestine' in *Palestine Exploration Fund Quarterly Statement*, London, January 1879–1880.
Firestone, Ya'akov, 'Crop-sharing economics in mandatory Palestine', *International Journal of Middle East Studies*, vol. II, no. 1, January 1975.

Fish, Henry C., *Bible Lands Illustrated*, Connecticut, 1876.
Glaisher, James, F. R. S., *Meteorological Observations at Jerusalem*, Palestine Exploration Fund, London, n.d. but after 1901.
Harding, G. Lancaster, *The Antiquities of Jordan*, London, 1967.
Heyd, Uriel,*Ottoman Documents on Palestine*, Oxford, 1960.
Hill, Gray, *With the Beduins*, London, 1891.
Himadeh, Sa'id, 'Taxation in the 1900s', in Charles Issawi, ed., *The Economic History of the Middle East 1800–1914*, Chicago, 1966.
Hoade, Fr Eugene, *East of the Jordan*, Jerusalem, 1954.
Hourani, A. H., *Syria and Lebanon*, Oxford, 1946.
Hunder, W. P., *Narrative of the Late Expedition under the Command of the Hon. Sir Robert Stopford*, London, 1842.
Hütteroth, Wolf-Dieter, 'The fiscal administration in the areas bordering on the desert', in *Second Conference, History of Bilad al-Sham*, Damascus, 1978.
Hütteroth, Wolf-Dieter, and Abdulfattah, Kamal, *Historical Geography of Palestine, Transjordan and Southern Syria in the Late 16th Century*, Erlangen, 1977.
Ibrahim, Dr M. M., 'Archeological Excavations at Sahab 1972', in Yosef Jamal Alami ed., *Annual of the Department of Antiquities*, no. 17, Amman, 1972.
Ionides, M. J., *Report on the Water Resources of Transjordan and their Development*, London, 1939.
Irby, The Hon. Charles Leonard and Mangles, James, *Travels in Egypt and Nubia, Syria and Asia Minor*, London, 1823.
Issawi, Charles, ed., *The Economic History of the Middle East*, Chicago, 1966.
Jaussen, P. Antonin, *Coutumes des arabes au pays de moab*, Paris, 1948.
Jarvis, Major C. S., *Arab Command*, London, 1942.
Karpat, Kemal H., 'Land regime, social structure and modernization in the Ottoman Empire', in W. R. Polk and R. L. Chambers eds, *Beginnings of Modernization in the Middle East*, Chicago, 1968.
Karpat, Kemal H., 'Some historical and methodological considerations concerning social stratification in the Middle East', in C. A. O. Van Nieuwenhuijze, ed., *Commoners Climbers and Notables*, Leiden, 1977.
Klein, Revd F. A., 'Missionary tour into a portion of the Transjordanic countries', pamphlet published by Church Missionary Society, April, 1869.
Lambton, A. K. S., *Landlord and Peasant in Persia*, London, 1953.
Lawrence, T. E., *Seven Pillars of Wisdom*, London, 1935.
Lewis, Bernard, 'Studies in the Ottoman archives', *Bulletin of the School of Oriental and African Studies*, University of London, vol. XVI, 1954.
Libbey, William and Hoskins, Franklin E., *The Jordan Valley and Petra*, 2 vols, New York, 1905.
Luke, H. C. and Keith-Roach, E. eds, *The Handbook of Palestine and Transjordan*, Jerusalem, 1930.

Lynch, W. F., *Narrative of the United States Expedition to the River Jordan and the Dead Sea*, Philadelphia, 1850.
Mantran, Robert, and Sauvaget, Jean, *Règlements fiscaux ottomanes. Les provinces syriennes*, Paris, 1951.
Ma'oz, Moshe, *Ottoman Reform in Syria and Palestine 1840–1861*, Oxford, 1968.
Ma'oz, Moshe, ed., *Studies on Palestine during the Ottoman Period*, Jerusalem, 1975.
Mavromatis, E., 'Preliminary scheme for the irrigation of the Jordan Valley', pamphlet, London, October 1922.
Medebielle, Pierre, *Salt – histoire d'une mission*, Jerusalem, 1955.
Medebielle, Père Pierre, 'La difficile installation', in *Histoire de Madaba, Le bulletin diocesain du patriarchat Latin*, no. 103, Jerusalem, 1985.
Merrill, Selah, *East of the Jordan*, London, 1881.
Moorman, F., 'The soils of Jordan', Expanded Technical Assistance Program, F.A.O. no. 1132, Rome, 1959.
Mufti, Shawkat, *Heroes and Emperors in Circassian History*, Beirut, 1972.
Musil, Alois, *Arabia Petraea*, 3 vols, Vienna, 1908; English version, New York, 1927.
Musil, Alois, *The Northern Hegaz*, New York, 1926.
Napier, Commodore Sir Charles, *The War in Syria*, London, 1842.
Napier, Lieut. Col. E., *Reminiscences of Syria*, London, 1843.
Neil, Revd James, 'Pits in the Shittim Plain' in *Palestine Exploration Fund Quarterly Statement*, 1891.
Nelson, Cynthia ed., *The Desert and the Sown. Nomads in a Wider Society*, Berkeley, 1973.
Oliphant, Laurence, *The Land of Gilead*, London, 1880.
Palestine Exploration Fund Quarterly Statement, 1868–9, 1891.
Partington, J. R., *Alkali Industry*, London, 1919.
Peake, F. G., *History and Tribes of Jordan*. Coral Gables, Fla., 1958; see also under Peake in Arabic section of bibliography.
Pococke, Richard, *Description of the East and Some Other Countries*, 2 vols, London, 1743–5.
Poliak, A. N., *Feudalism in Egypt, Syria, Palestine and the Lebanon 1250–1900*, London, 1939.
Ponsonby, A., *Handbook of Arabia*, vol. I, General, HMSO (no. 1.D.1128), Oxford, probably 1920.
Possetto, Alessandro, *Il patriarcato latino de Jerusalemme 1848–1898*, Milan, 1938.
Rafeq, Abdul-Karim, *The Province of Damascus 1723–1783*, Beirut, 1970.
Robinson Lees, Revd G., *Life and Adventure beyond Jordan*, London, probably 1908.
Rogers, M. E., 'Maritime cities and plains in Palestine', in Charles Wilson ed., *Picturesque Palestine*, vol. 3, London, probably 1880.
Roseman, N., 'One hundred years of rainfall in Jerusalem', *Israel Exploration Journal*, vol. 5, Jerusalem, 1955.

Russell, Josiah C., 'The population of medieval Egypt', *Journal of the American Research Center in Egypt*, vol. 5, Cairo, 1966.
Savignac, Fr M. R., 'Une église byzantine', *Revue Biblique*, no. 12, Paris, 1903.
Schaff, Phillip and Rogers, M. E., 'Damascus', in Charles Wilson ed., *Picturesque Palestine*, London, probably 1880.
Schumacher, G., *The Jaulan*, London, 1888.
Schumacher, G., *Northern Ajlun*, London, 1890.
Schumacher, G., *Across the Jordan*, London, 1886.
Seetzen, M., *Brief Account of the Countries Adjoining the Lake of Tiberias, The Jordan and the Dead Sea*, The Palestine Association of London, London, 1810.
Seikaly, May, 'The Arab community of Haifa 1918–1936', unpublished D.Phil. thesis, Oxford, 1983.
Seteny, Shami, 'Anthropological research and the local community', unpublished paper presented to the Symposium on Anthropology in Jordan, Amman, 25–28 February 1984.
Shahid, Irfan, *Byzantium and the Arabs in the Fourth Century*, Washington, 1984.
Smith, C. Gordon, 'The geography and natural resources of Palestine as seen by British writers in the nineteenth century and early twentieth century', in Moshe Ma'oz ed., *Studies on Palestine during the Ottoman Period*, Jerusalem, 1975.
Spyridon, S. N., 'Annals of Palestine 1821–1841', in *Journal of the Palestine Oriental Society* 17 (1938).
Tash, Waleed, 'The Circassians in the Middle East', paper presented to the seminar 'Minorities in the Arab East and Israeli Attempts for Manipulation', Amman, 12–15 September, 1981.
Thomson, William M., *The Land and the Book*, London, 1881.
Tooley, R. V., *Maps and Map-Makers*, London, 1982.
Tristram, H. B., *The Land of Moab*, London, 1873.
Tristram, H. B., *The Land of Israel*, London, 1865.
Tute, R. C., *The Ottoman Land Laws*, Jerusalem, 1927.
Van Lennep, Henry J., *Bible Lands: their Modern Customs and Manners*, London, 1975.
Volney, M. C. F., *Travels through Syria and Egypt*, trans. C. G. J. and J. Robinson, 2 vols, London, 1787.
Walsh, W. Pakenham, *The Moabite Stone*, Dublin, 1873.
Waterfield, Gordon, *Layard of Ninevah*, London, 1963.
Wavell, A. P., *The Palestine Campaigns*, London, 1928.
Weightman, G. H., 'The Circassians', in Charles W. Churchill and Abdulla M. Lutfiyya eds, *Readings in Arab Middle Eastern Societies and Cultures*, New York, 1970.
Wilkinson, Frederick, *The World's Great Guns*, London, 1977.
Wilson, Charles ed., *Picturesque Palestine*, 3 vols, London, probably 1880.

Glossary

'adaniya (coffee beans)
'afir (term applied to areas sown before the rains)
aghnam (tax levied on animals)
aman (safe conduct)
ankari (large copper plates able to accommodate five or six men each)
'aqd (Roman vault)
'araq (alcoholic drink distilled from grapes)
ardiya (demurrage)
'ashabin (weeders)
a'shar (taxes on crops)
'askariya (military tax)
'aysh (crushed wheat boiled in water)

badal a'shar (in lieu of the tithes)
badayil (exchange)
bathar (seed thrower)
batiha (measurement of 48 *sa's*)
bawwab(doorkeeper)
bayadir (pl. of *baydar*, q.v.)
bayadiri (baydar supervisor)
baydar (threshing or threshing ground)
bayika (pl. *bawayik*, stable)
baylik (the farm's household)
bayt al-sha'ar (tent)
bidhar (seeds)
bidhar dallil (sparse seeding)

bidhar i 'ba (intensive seeding)
buyut sha'ar (black tents)

dabka (peasant dance)
daftardar (chief financial officer of a province)
daftars (registers)
darras (thresher)
dawla (government)
dharra (sifting)
dhurra (millet)
dibs (syrup made from juice of grapes or dates)
dira (tribal domain)
diras (threshing)
dunums (see appendix 2)

faddan (see appendix 2)
fallahin (farmers or peasants)
faqqus (a variety of cucumber)
farah (joy)
farik (roasted green wheat)
ferde (personal tax)
fitaha al-manjal (the opening by the sickle)
futuh al-ard (the opening of the land)

ghalath (contaminated seeds)
ghamarin (bundle gatherers)
gharbala (sifting)
ghariba (strange seeds)
ghazu (raid)
ghirbal *(sieve)*
ghrara (weight of nearly 750 kilograms)
ghubra al-baydar (*baydar* dust)
ghumr (bundle)

habb al-hal (cardamom seeds)
halat (lit. unwind: the expression for rest when the rabta stopped work)
halawa (sweetmeat made from sesame oil and sugar)
hamayil (clans)
haqla (harvest field)

haqq al-qarar (claims of ownership)
harathin (ploughmen)
hasida (using the sickle – harvest)
hassadin (harvesters)
hatta (*kufiya*, the Arab head-dress, worn with the *‘iqal*)
hawra (leather apron)
hulba (fenugreek)

i‘ana akhshab (wood aid)
‘idal (bag made out of woven wool cloth)
iltizam (tax contract; tax farm)
‘iqal (headband)
irada (imperial order)
‘ishaba (weeding)
‘izbiya (women cooks-bakers)

jahush (lit. donkey-rider: term used to refer to an older harvester at the end of the line during harvesting)
jamid (dried yoghurt)
jammal (camel man)
jarisha (crushed wheat)
jarra (jar)
jawra‘a (feast celebrating the end of the harvest)
jazur (she-camel specially fattened)
jiftlik (imperial domain)

kabra (additional share as a sign of reverence and seniority)
kadis al-qamh (wheat haystack)
kadish (pl. *kudsh*, castrated horse)
karadish dhurra (loaves made from millet)
kasr al-ard (breaking of the soil, to cultivate the land)
katib (scribe or secretary)
kayl (see appendix 2)
kayyal (he who measures cereals)
khanat (households)
khashuqa (wooden spoons for cooking)
khatm (wooden seal)
khawa (tribute or protection money)
khaysha (pl. *khiyash*, huge bag made of jute)
khazina ‘amira (treasury)

khirba (pl. *khirab*, deserted village)
khiyash (pl. of *khaysha*. q.v.)
khubz tabun (ordinary bread)
khubz shrak (thin bread)
kirab (fields deeply ploughed)
kirbal (implement used for the sifting and carrying of straw)
kudsh (sing. *kadish*, q.v.)

laban (yoghurt or sour milk)
labna (yoghurt or yoghurt that has been strained through a cloth bag)
lahham (butcher)
laqatat (gleaners)
lawh diras (threshing board)
liwa‘ (sub-province or sanjaq)
luzayqiyat (special dish made from khubz shrak covered with butter and sugar)

madafa (guest house)
madhari (prongs)
mahkama shar‘iya (Islamic law court)
mahr (dowry or marriage portion)
majlis da‘awa (juridical council or court)
majlis idara (administrative council)
maktubji (secretary)
mal (money or tax)
malliq (term for ploughed fields planted without rest)
manajil (sickles)
manaqil (ember containers)
manqal (pl. *manaqil*, fire container)
mansaf (large platter)
maqathi (summer vegetble fields)
maqatir (pl. of *maqtar*, q.v.)
maqsam akhu (share of a brother)
maqtar (pl. *maqatir*, camel caravan)
masarif filaha (farming expenses)
mashyakha (pl. *mashyakhat*, chieftainship)
midhra (pitchfork)
miqra‘a (a crude whip)

miri (tax; that which pertains to the amir – *miri* land was land owned by the state)
mu'allim (pl. *mu'allimin*, master, lit. teacher; landlord)
mudd (measure of 3 *sa's* or 18 kilograms of wheat)
mudir (governor)
mudir mal (fiscal officer)
mudir muhasaba (fiscal director)
muhasibji (fiscal director)
mukhtar (headman)
mulk (of land: possessed in full ownership; private)
multazim (tax-farmer)
muna (supplies, provisions)
muqasama (system of the division among parties of a share from the crop)
muraba'a (division into four; sharing of crop whereby labour receives a quarter)
murabi' (labourer)
murabi'iya (farm hands on yearly contract)
musha' (common land-ownership system)
mutasallim (governor)
mutasarrif (governor of a sanjaq)
muzara'a (farming arrangements)

nada (dew)
nahiya (sing. *nawahi*, sub-district)
najis (dirty, unclean)
narghila (hubble-bubble)
nawahi (sing. of *nahiya* q.v.)
nizam al-iltizam (the system of tax-farming)
nizam-i cedid (new troops)
nuqra (fire hearth)

qa'immaqam (governor of a *qada')*
qada' (district)
qahwa al-sada (black coffee)
qahwaji (coffee maker)
qasal (straw)
qasida (poem)
qatruz (pl. *qatariz*, stable boy)
qila'a (pulling the plants from the roots)

qilw (soda ash)
qufra (fat)
qutu‘ (built-in compartments for the storage of seeds)

rabta (pl. *rabtat*, work gang)
rajjad (camel-driver)
ramy al-bidhar (broadcasting of seeds)
rashm (wooden seal)
riy (term applied to areas sown after the rains)
rub‘ (quarter; quarter allotted for the labour force)
rumman (pomegranates)

sa‘ (measurement equivalent to 6 kilograms)
sabiya (riding contest)
sahja (beduin dance)
Salname (official yearbook)
salyane (annual tax)
samir (party)
samna (melted butter or Syrian butter ghee)
sanad (sale document)
sanjaq (sub-province)
saqqa (water carrier)
sarr (purses)
sawama (very large farm owning large numbers of animals for different kinds of work)
serai (government house)
shaddad (pl. *shaddadin*, net handler)
shaquq (lit. the one who cuts: the head of the *rabta* during harvesting)
sharaha (gluttony)
sharaka (pl. *sharakat*, partnership)
sharha (name given to the gifts of land given to shaykhs of tribes)
shawa‘ib (pl. *sha‘awb*, hand tool to turn the hay during threshing)
shaykh (pl. *shuyukh*, chief or landlord)
shaykh al-mashayikh (paramount chief)
shibaba (flute)
shiqaq (operation opening new land to cultivation)
shuyukh al-rabtat (*rabta* chiefs or foremen)

sinsla (stone wall)
siraskir (army commander)
siyaq (common term for dowry or marriage portion)
subba (the name given to the heap of seeds at Baydar which had been sifted)
sukhra (forced, unpaid labour)
summaq (sumach)
surra (purse)

tabun (baking oven)
tafwid (transaction allocating the use of land)
tahna (wheat milled for flour)
tapu (land registry)
tarbush (red fez used during Ottoman times by city dwellers)
tarha (heap, hay batch)
tassa (mug)
thilath (the third ploughing of a field)
thnaya (the second ploughing of a field)
tibn (hay)

'udul (sacks)
'ushr (one-tenth tax, or tithe)

wadi (river valley)
wajh (face or line of the harvesters in the field)
wakil (steward)
wali (governor)
waqf (endowment)
waqqaf (overseer)
werko (also written *vergo*, property tax)
widy (tax on animals)
wilaya (pl. *wilayat*, province, governorate, governorship)

yiktallu (to buy by the measure)

al-zagharit (shrill trills of joy)
zahfa (crawling)
zaptieth (irregular troops)

Index